T0258659

IEE ELECTROMAGNETIC WAVES SERIES 12

Series Editors: Professors P. J. B. Clarricoats
E. D. R. Shearman and J. R. Wait

MICROSTRIP ANTENNA

THEORY
AND
DESIGN

Other volumes in this series

MICROSTRIP
ANTENNA
THEORY
AND
DESIGN

J. R. JAMES,
P. S. HALL
and
C. WOOD

Peter Peregrinus Ltd on behalf of the Institution of Electrical Engineers

Published by: Peter Peregrinus Ltd, London, United Kingdom
© 1981: Peter Peregrinus Ltd

Paperback reprint 1986

ISBN 0 86341 088 X

Contents

Preface

In the past few years, the concept of creating microwave antennas using microstrip has attracted increasing attention and viable practical designs are now emerging. The purpose of this monograph is to present the reader with an appreciation of the underlying physical action, up-to-date theoretical treatments, useful antenna design approaches and the overall state-of-the-art situation. The emphasis is on antenna engineering design but to achieve this goal it has been necessary to delve into the behaviour of microstrip in a much wider sense and also include aspects of electromagnetic analysis. As a consequence, the monograph will also be of interest to microstrip circuit designers and to some extent those seeking electromagnetic problems of a challenging nature.

It is worthwhile saying a little about the trends that led up to these new types of antennas. To begin with, the idea of utilising a printed conductor on a substrate as a radiating element is probably as old as the idea of printed transmission lines themselves. Microstrip transmission lines were introduced in the 1950s and it was evident then that the lines, together with their compatible microstrip components, were prone to radiation losses and parasitic coupling effects which became more troublesome as the operating frequency increased. It is interesting to reflect that the antenna designer actually makes good use of, and optimises, one of these unwanted properties, the radiation. One may well wonder why the research into microstrip antennas did not intensify at an earlier date and there are perhaps several reasons that can be pointed out. The astronomical progress in miniaturising and integrating electronic circuits in the past decade has recently created a positive demand for a new generation of antenna systems. In principle, microstrip antennas are thin planar configurations that are lightweight, low cost, easy to manufacture and can be made conformal with the surfaces of vehicles, missiles etc. The compatibility of microstrip antennas with integrated electronics is thus very evident and is a great impetus to antenna designers particularly so, now that many new types of substrate materials are commercially available. However, the microstrip wave-trapping effects, so vital for circuit operation, inhibit the radiation mechanism and must be taken into account in antenna design. This seems an obvious statement when one considers that the original intended purpose of the microstrip structure was to act as an efficient transmission line. Obvious though this may seem, the earlier attempts to make microstrip antennas were based on the expectations that a half-wavelength printed conductor should radiate as freely as its free-space counterpart. Clearly, the wave-trapping effects of the microstrip line and surrounding substrate must be embraced in any physical model and this has been a prominent feature of our own research at RMCS from the start, enabling us to evolve many novel antenna designs. Wave trapping effects in substrates involve the study of surface waves and discontinuities in open waveguide structures; these topic areas are in themselves not fully expoited and exact mathematical analysis can only be carried out for very simple geometries and this at present excludes microstrip discontinuities. Analytical treatments of arrays of microstrip radiators are therefore only likely to be fruitful through careful approximation based on physical reasoning. It is clear that the microstrip antenna designer must encompass many more effects than previously considered by microstrip circuit designers. It is for these reasons that the scope of this monograph is necessarily somewhat wider than the title may suggest.

The first Chapter introduces other types of planar antennas that have been developed in previous years and highlights their advantages and disadvantages. Although this background review is of interest in itself, it identifies microstrip antennas as the latest attempt to evolve thin planar low-cost structures. When viewed in this context many of properties and problems can be anticipated, and set the scene for subsequent Chapters. An early opportunity is taken in Chapter 2 to collate all the relevant well known properties of microstrip lines together with useful formula prior to introducing the more recent treatment of radiation mechanisms of microstrip discontinuities in Chapter 3. Various ways of measuring these effects are also included and recommended formulas and data are listed. This leads logically to the understanding and design of the most elementary type of microstrip radiator, the resonant patch in Chapter 4. More complicated linear arrays are dealt with in Chapter 5 but their stacked use in two-dimensional arrays, Chapter 6, necessarily includes a detailed treatment of corporate feed design which is a topic in itself. Chapter 7 presents interesting developments in microstrip antennas that have circularly polarised characteristics. Chapter 8 highlights some of the manufacturing problems that have been recently encountered by designers whereas in contrast, Chapter 9 deals with recent advances in the mathematical analysis of microstrip antennas. In the final Chapter 10, current design trends are identified and impressions extracted as to what the future holds for microstrip antennas. Some supporting details are given in Appendixes.

In the interests of clarity, each Chapter concludes with a summary comment on its content together with a list of references. As far as we are aware no other comparable text on microstrip antennas is available at the time of going to press and we hope that readers will find our book of some value to them in their endeavours in this fascinating modern class of antenna.

November 1980

J.R. JAMES
P.S. HALL
C. WOOD

Acknowledgments

We warmly thank all our colleagues who have contributed in some way or another to the writing of this book. In particular we are very grateful to Dr. Ann Henderson and Mr. Eric England of the Royal Military College of Science for helpful discussions and contributions to our research work on microstrip antennas, to Dr. Nick Taylor of the Radar and Signals Research Establishment, Malvern, who was involved with us on several Ministry of Defence contracts on microstrip antennas and to Decca Radar Ltd. for some information on substrates given in Chapter 8. We would like to acknowledge the encouragement we have received from the Dean of the College, Dr. Francis Farley, F.R.S., by way of his interest in our research, helpful discussions and permission to publish this book.

Principal symbols

β, β_s = microstrip and substrate phase constants; also spectral terms β_r

ω = angular frequency

h = substrate height

w, w_e = conductor width and effective width

ϵ_0, ϵ_r = free space and relative permittivity

μ_0, μ_r = free space and relative permeability

ϵ_e = effective relative permittivity

λ_0 $\begin{cases} = \text{free space wavelength} \\ = \text{wavelength in air-spaced microstrip} \end{cases}$

$\lambda_m, \lambda_s, \lambda_g$ = wavelength in microstrip, substrate and waveguide

$\tan \delta$ = loss tangent of dielectric

σ_c, σ_d = conductivity of conductors and dielectrics

Z = free space wave impedance = $(\mu_0/\epsilon_0)^{1/2}$

Z_m = impedance of microstrip = $(1/Y_m)$

Z_0, Z_d = impedance of air-spaced microstrip and microstrip completely surrounded by dielectric material

k_0 = $2\pi/\lambda_0$

VSWR   = voltage standing wave ratio

Chapter 1

Flat-plate antenna techniques and constraints on performance

Microstrip is the name given to a type of open waveguiding structure that is now commonly used in present-day electronics, not only as a transmission line but for circuit components such as filters, couplers, resonators etc. Readers not completely familiar with the microstrip concept are referred to Chapter 2 for some explanatory details and references. The idea of using microstrip to construct antennas is a much more recent development and an example of a microstrip antenna is given in Fig. 1.1. The antenna assembly is physically very simple and flat which are two of the reasons for the great interest in this new antenna topic. The upper surface of the dielectric substrate supports the printed conducting strip which is suitably contoured while the entire lower surface of the substrate is backed by the conducting ground plate. Sometimes the antennas are referred to as printed antennas because of the manufacturing process.

Many new types of microstrip antennas have evolved which are variants of the basic sandwich structure but the underlying concept is the same; that is the printed radiating element is electrically driven with respect to the ground plane. This definition admits a wide range of dielectric substrate thicknesses and permittivities and of course the special case when the strip and ground are separated only by an air space. This then is what is generally understood by the term microstrip antenna.

The concept of creating antennas from printed microstrip circuits provides the antenna engineer with an opportunity to construct very thin planar antenna structures. The performance obtainable will of course relate to the specific requirements but there are general characteristics which can be identified; this in turn gives a useful indication of the performance of microstrip antennas that can reasonably be anticipated. This is particularly so for such parameters as antenna bandwidth and efficiency whereby an estimate is often sufficient to judge the feasibility of a design specification. In order to get an appreciation of the potential of microstrip antennas it is expedient to examine other types of flat-plate antennas and their limitations. In this context, the microstrip antenna does not appear as a device in isolation but rather the latest development in the pursuit of low-cost very thin flat antenna structures. Looked at in this way this new microstrip antenna technology does not hold too many surprises.

Fig. 1.1   *Microstrip antenna employing four short-circuit patch elements, fed to radiate circular polarisation*
The two centre elements are capacitively coupled (See details in Hall, Wood and James, 1981)

The extreme miniaturisation of electronic components in recent years often leaves the antenna as the most bulky device in the equipment; the demand for smaller, lightweight antenna systems is thus evident. A commonly occurring requirement is for antennas with flat profiles that can be made flush with walls of vehicle and aircraft bodies or, alternatively, are ideally situated to manpack microwave systems. Typical applications include intruder alarms, aircraft radioaltimeters, portable radar, microwave telecommunication and telemetry links; more recently lightweight flat arrays are being considered for missile-guidance systems. Ways of making antenna arrays flat and thin have featured strongly as a research area embracing both conventional waveguide devices and the newer printed assemblies. Many of the conventional slotted waveguide array techniques are well established and it is not intended here to give a comprehensive coverage of all these methods but rather to extract the fundamental principles and limitations of the various array types to illustrate their logical development. The types can be conveniently subdivided into those antennas employing waveguide, triplate lines, printed cavity backed assemblies and finally microstrip, as follows.

1.1 Waveguide antennas

Anyone contemplating the design of some new form of planar array should first get acquainted with waveguide array configurations (Jasik, 1961) because these are generic forms that are common to a wide variety of other antenna structures including the more recent printed versions. A suitably oriented slotted hole in a waveguide radiates and a linear array can be formed by placing several slots periodically along a waveguide run. The slots radiate power from the incident waveguide mode which may then be reflected by a terminal short circuit to create a narrowband resonant array. Alternatively, if the residue of the incident wave is absorbed by a load, then a broadband travelling-wave array is produced. If the power radiated by the elements has a shaped distribution across the array then the sidelobes can be reduced. Waveguide flat profile arrays consist of several slotted waveguide sections placed in parallel and fed by either a branched system of waveguides, or a more compact series waveguide feed. The main beam of a resonant waveguide array is normal to the array, whereas the beam of a travelling-wave array is designed to squint from normal and this squint varies with frequency. These conditions are common to all series-fed arrays, whereby power enters the array at one end and is progressively 'tapped off' and radiated as it progresses towards the other end. The fact that each radiating element does not have to be separately excited is the major advantage obtained at the expense of problems with precise element control in relation to matching, sidelobe, bandwidth and squint characteristics.

1.1.1 Slot element design and deployment
The slot elements impose a small complex load on the waveguide and some basic slot arrangements are shown in Fig. 1.2(*a*) and (*b*) together with their equivalent circuits. Stevenson (1948) derived the equivalent R and G values using Babinet's

principle assuming very thin waveguide walls. Oliner (1957) extended the treatment to obtain Z and Y for finite wall thickness. Measurements by Cullen and Goward (1946) and Watson (1946) confirm these expressions but mutual coupling can be significant and has to be allowed for (Harvey, 1963a). Ideally, the slot element needs to be tuned, incur minimal loss, have small size, radiate with the desired polarisation, provide the desired degree of coupling to the guide, have adequate bandwidth and when also used as a terminal load it must present a match to the waveguide. These are generally conflicting requirements and a compromise solution is sought. Some of the techniques used are well worth noting because they exhibit fundamental properties. For instance, the slot size can be reduced by dielectric

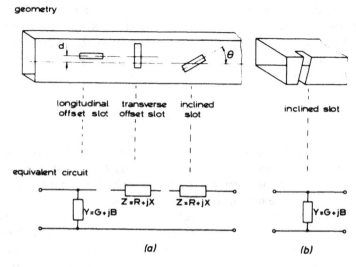

Fig. 1.2   *Slot configuration and equivalent circuit for slotted waveguide antennas*
   a Broad wall slots
   b Narrow wall slots

loading but then the bandwidth of the slot is reduced, and for high permittivity material the loss in the latter can be a consideration; similar effects have been analysed by James and Henderson (1978a) in relation to small dielectric coated monopoles. The amount of power coupled to the slots needs to be under the designers control, since the available power level decreases progressively along the waveguide run and in addition some form of amplitude taper across the linear array will almost certainly be required to achieve a desired sidelobe level. This degree of coupling may vary between -5 and -40 dB and is obtained by varying the slot position in the waveguide and using different shaped slots in the form of a letter T or H or I, (Oliner and Malech, 1966a). New types of elements are still being evolved, but there is little scope for altering the basic nature of the slot pattern which is essentially that of a Hertzian magnetic dipole element embedded in a conducting plane. However, Clavin *et al.* (1974) employ a combination of parasitic wires above the slot to equalise the principal plane radiation patterns and reduce mutual coupling.

A crossed-slot arrangement can be employed to act as two orthogonal dipoles and hence with suitable phasing provide circular polarisation. Many other configurations have been reported including a fed dipole in a parasitic slot (Cox and Rupp, 1970) and more recently a combination of separate transverse and longitudinal slots in a resonant array with selective mode feeds, (Palumbo *et al.*, 1973). Arranging a pair of orthogonally polarised elements in close proximity and obtaining the desired excitation phase is therefore the main problem to solve with circularly polarised antennas, and bandwidth constraints are likely to arise.

1.1.2 Waveguide arrays

An outline sketch of a two-dimensional travelling-wave array composed of several linear arrays, is sketched in Fig. 1.3. The need for a well designed feeding arrangement is evident and the individual linear arrays can be fed in series as in Fig. 1.3 or in parallel. In the latter case, a system of branch lines called a corporate feed connects the arrays to a single feed point.

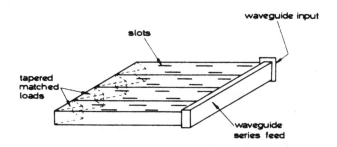

Fig. 1.3   *Typical waveguide flat-profile antenna array using slots or leaky wave action*

For resonant arrays, the radiating elements are normally placed $\lambda_g/2$ apart and are arranged to produce alternately antiphase radiation; λ_g is the wavelength in the guide. Variation of the individual slot coupling allows the correct distribution to be obtained and the result (Silver, 1965a) is a very narrow bandwidth, input-impedance-controlled array, with a bandwidth of typically less than 1% but with a broadside beam. The attraction of the resonant array is therefore its simplicity and lack of squint but the resulting bandwidth is very narrow (Harvey, 1963b). In a travelling-wave array, the wider bandwidth is obtained by spacing the radiating elements slightly differently from $\lambda_g/2$ so that in a long array, a stable cycle is set up in which the reflected waves from the slot conductances are essentially cancelled by the effect of the spacing. Experimental results confirm that a VSWR of about 1·2 can be achieved for a spacing greater than $0.55\,\lambda_g$. About 5% of the total power is typically absorbed in the matched load (Silver, 1965b); this matched load removes the 'ghost beam', due to reflected power, on the opposite side of the normal to the main beam. However, in this type of array the beam squints with frequency which is a definite disadvantage in most applications. It can be used to advantage as a frequency scanned antenna whereby the beam position can be rapidly shifted according to the operating frequency. Craven and Hockham (1975) have used evanescent waveguide sections to form a frequency-scanned planar linear array, with a scan rate of about 16° per cent frequency change being achieved in one case. The opposite problem of creating an array that is insensitive to frequency can in principle be solved by ensuring an equal path length to each element. This leads to a bulky corporate feed system. A more compact form of squintless array, shown in Fig 1.4 has been developed by Rodgers (1972). Mutual coupling between elements can be a problem in large arrays having many elements where there is a requirement to scan over a wide range of angles. Looked at in another way, the array

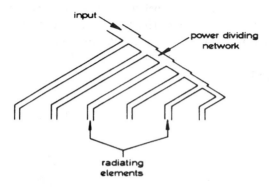

Fig. 1.4   *Feed arrangement for squintless array*

appears as a periodic structure and it is easy to appreciate that the latter can act as a wave supporting device in certain pass bands. This gives rise to the well known blindness effects, (Knittel, 1972; Oliner and Malech, 1966b). Mutual coupling also creates matching problems due to impedance changes. On the whole, effects due to mutual coupling are both difficult to calculate and measure due to the multiplicity of elements in a large array and the geometry of each element. In the slotted waveguide array, the problem is somewhat simplified because the element geometry and feed arrangement is well defined as a boundary value problem. It is also possible to have one continuous slot along a waveguide to create a leaky-wave antenna but the main beam direction is confined to between 10° and 80° to the normal. (Oliner and Malech, 1966b).

1.1.3   Dielectric waveguide arrays
The recent requirements for millimetre wave radar equipment has stimulated interest in the design of new types of antennas for this frequency range where metal waveguides exhibit unacceptable losses. New types of dielectric waveguide structures have been investigated which also allow some degree of beam scanning, (Itoh and Hebert, 1978; Klohn *et al.*. 1978). Planar sandwich antennas have also

been demonstrated at submillimeter wavelengths by Hwang, Rutledge and Schwarz (1979). These are very new techniques which will no doubt result in much innovation.

1.2 Triplate antennas

A triplate transmission line, often referred to as stripline, consists of a flat conducting strip suspended in a symmetrical position between two parallel ground planes. The structure is usually made using two sheets of copper-clad dielectric and the central conductor is printed on to one of the substrates prior to bonding the two dielectric surfaces together. Various other physical arrangements exist and for the transmission of greater power, air spacing is employed. The guided wave is confined between the ground plane in the vicinity of the central conductor and the basic mode is a transverse electric magnetic (TEM) wave, (Cohn 1954). Triplate is a thinner, lighter weight and lower-cost guiding structure than conventional waveguide and it can be made to radiate in a similar fashion by creating slots in one of the ground planes.

1.2.1 Design of the slot element

Oliner (1954) derived an approximate expression for the conductance of a slot in a ground plane fed by a symmetrically placed strip and Breithaupt (1968) extended this to include asymmetrically fed slots. Offsetting the slot from the symmetric position reduces the amount of power coupled to the slot. An accurate calculation of the slot susceptance is difficult and one reason is the fact that the presence of the slot disturbs the symmetry of the triplate line and allows higher-order modes to be excited in the latter; in particular the so-called parallel-plate mode. Rao and Das (1978) have recently addressed the problems. In practice this higher-mode coupling is very undesirable if several slots are to be fed from one triplate line and one technique is to place metal pins around the slot as shown in Fig. 1.5. This makes the slot element design reliant on trial and error methods because the precise action is difficult to assess but if sufficient pins are used they can be regarded as forming the walls of a cavity. The susceptance of cavity backed rectangular slots has been analysed by Galejs (1963), Adams (1967) and Lagerlof (1973). This has provided useful quantitative data on the behaviour of slot elements and indicated that the loading of the slot cavity with dielectric reduced the bandwidth.

1.2.2 Triplate slotted arrays

Despite the empirical nature of design methods for triplate slot elements, several useful antenna arrays have been developed using dielectric filled triplate structures. Offset feed lines have been used by Hill and Paul (1975) in a five-element linear array having a − 18 dB sidelobe level and a VSWR < 1·3 over a 4% bandwidth. Using a similar feed arrangement, shown diagramatically in Fig. 1.6, Bazire *et al.* (1975) have produced an array with four offset beams independently accessible from each feed point and this array is used in Doppler radar navigation systems. The empirical design was based on an analysis of Galejs (1962). The final design consisted of a 9 × 27 element array and the cosine-squared *E*-plane distribution resulted in sidelobes typically below − 25 dB. The array had a VSWR < 1·2 with line losses of 0·75 dB.

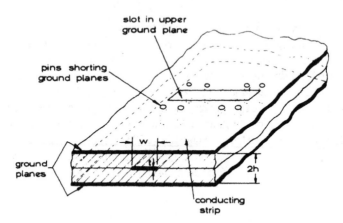

Fig. 1.5 *Stripline slot radiator and feed arrangement showing shorting pins*

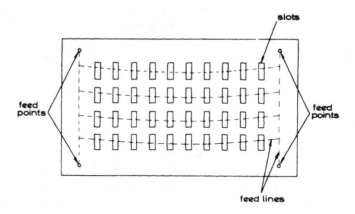

Fig. 1.6 *Array of stripline slot radiators with four feed points offering a choice of beams*

An array having 52 elements is described by Josefson *et al.* (1974) having a highest sidelobe level of − 25 dB and the VSWR < 2·5 over a 6% bandwidth with line losses of about 2 dB. Fritz and Mayes (1974) have made a frequency-scanned array using a stripline slot which has been made to scan through broadside without producing a large input VSWR by employing matched elements. Campbell (1969) has shown that the stripline slot can operate with a dielectric layer over it, thus affording improved environmental performance. Another way of creating a cavity effect behind the slot is to surround the slot with 'through-plated' holes; various multi-sandwich triplate assemblies have also been developed. With all these triplate assemblies the sidelobe levels obtainable are ultimately constrained by production tolerances which are generally more critical for the more complicated structures. Sidelobe levels no better than $-25\,dB$ appear to be the state-of-the-art for high directivity arrays above 10 GHz. An entirely different slot structure has been explored by Laursen (1973) who used one long slot in the triplate ground plane and fed it by a centre conductor whose central position was periodically modulated. A practical array of eight such linear arrays at X-band had $-15\,dB$ H-plane sidelobe level, 3 dB worse than the design level and a VSWR $< 1\cdot4$ over a 5% bandwidth. Line losses were $0\cdot6\,dB$ and the loss in the terminating matched loads was $0\cdot7\,dB$. This structure does not appear to offer better sidelobe control than the individual slot arrays.

Finally, a feed arrangement that dispenses with symmetrically dielectric loaded triplate is the microstrip fed slot radiator Fig. 1.7, (Yoshimura, 1972) also described in Section 10.3.3. He showed the feasibility of creating a 4×4 X-band array

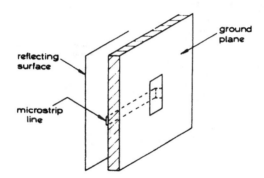

Fig. 1.7   *Ground plane slot radiator*

with a $-26\,dB$ sidelobe level and the input admittance is a function of the reflector to slot separation. Collier (1977) has described an array of 512 such elements; feed losses and sidelobe level control are seen to be severe constraints. Parasitic waves can clearly be excited in the region between the ground plane and reflector and it is unlikely that this asymmetrical triplate arrangement has an outstanding advantage over the symmetric triplate version Fig. 1.5. The analysis of the asymmetric case is certainly more involved.

1.3 Cavity-backed printed antennas

Much expertise on conventional wire antennas has accumulated over some fifty years, and it is well known that the wires need not be cylindrical but can be flattened into strips. The idea of printing dipoles as flat conducting strips on a substrate then follows logically. For many applications, directivity in the forward direction is the requirement and a conducting ground plane is included which is generally placed 0·25 wavelength behind the printed element. This concept has produced some very useful lightweight planar arrays and the bandwidths obtained are comparable with those of conventional wire antennas.

1.3.1 Printed dipole elements

There are now numerous versions of printed dipoles and two basic approaches are sketched in Fig 1.8. In Fig 1.8(*a*), each arm of the dipole together with its feed conductor is printed on each side of a thin dielectric substrate. The substrate has some detuning effect on the dipole which is consequently shorter than its freespace length.

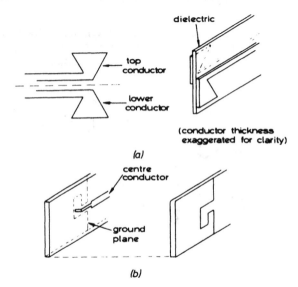

Fig. 1.8   *Printed dipole antennas*
a Dipole element
b Feed arrangements

The balanced feed arrangement can be a disadvantage and printed baluns have been devised, (Wilkinson, 1974). A dipole configuration that is compatible with triplate is sketched in Fig. 1.8(*b*), whereby the centre strip conductor protrudes beyond the two ground planes; the two arms of the dipole are then connected to the ground plane either side of the protruding strip. Hersch (1973) has described an array of two such elements having a VSWR < 1·5 over a 5% bandwidth. The deployment of 256 printed dipoles in an X-band uniformally illuminated array is also described by Wilkinson (1974). Losses in the elements and feed system amount to 3 dB and a VSWR < 1·3 over a 5% bandwidth is reported. Stark (1972) describes a similar array. Some earlier attempts at printing dipole structures included a version of the conventional Franklin antenna (Fubini *et al.*, 1955) whereby the phase reversing sections were also printed with the dipole elements, Fig 1.9(*a*); a 6 × 5 microwave array of these elements was found to be very narrowband. A capacitively coupled dipole array (McDonough, 1957) also appears to be very narrowband, Fig. 1.9(*b*). He also attempts to translate other wire antennas such as Yagi, Rhombic and Cigar antennas into printed versions. Much of the recent effort has been concentrated on evolving printed dipole elements that have a wide bandwidth and also have a smaller air gap between the substrate and ground plane to yield·a thinner structure. A successful design was created by Sidford (1975) and is illustrated in Fig. 1.10; a printed dipole is placed within a coplanar slot to produce

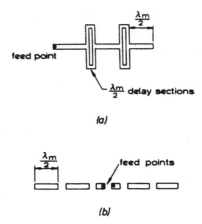

Fig. 1.9   *Printed colinear dipole arrays*
a Franklin antenna array
b Capacitively coupled array

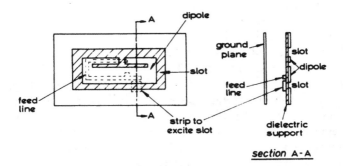

Fig. 1.10   *Flat hybrid arrangement of slot radiator and embedded dipole element*

circular polarisation. A bandwidth of 5% for a VSWR < 2 is obtained together with 2 dB ellipticity despite the small groundplane spacing of $\lambda_0/12$. A bandwidth of 21% with a VSWR < 2 has been reported by Dubost (1976) for a linear polarised configuration. These are described further in Section 10.2.3. Keen (1974) has demonstrated the feasibility of creating a printed log-periodic structure.

1.3.2 Cavity-backed printed arrays

When a large array of individually fed printed dipoles is constructed, the feeding networks present many problems involving unwanted radiation from bends etc. in the feed lines and the necessity for a bulky mechanical assembly. The need for a simpler, thinner feed arrangement is apparent and the travelling-wave principle has been adopted in the Sandwich wire antenna (Rotman and Karas, 1957). Individual printed dipoles are, however, not employed, and radiation occurs from a long periodically shaped printed conductor which is backed by a long rectangular cavity. This type of array construction has proved to be a useful concept which is also amenable to calculation (Green and Whitrow, 1971). Design improvements have been introduced by Graham *et al.* (1974) in the development of a monopulse array, Fig 1.11(*a*) allowing a bandwidth of 40% to be achieved and sidelobe levels of − 27 dB. A pillbox feed, Fig. 1.11(*b*), has been developed to feed the array to create a lightweight, rugged planar antenna with negligible power lost in the conductors and about 5% in the terminating loads.

Another novel travelling-wave arrangement is the Chain Antenna (Tiuri *et al.*, 1976) where the chain-like radiating element is a rectangular shaped meander line pair spaced about 0·1 wavelength from the ground plane, Fig 1.12(*a*). An interesting *L*-band array comprised of several such chain lines has been constructed as a net-like structure, Fig 1.12(*b*), but sidelobe levels reported so far are only − 10 dB. The chain antenna can also be regarded as comprised of air-spaced microstrip.

1.4 Microstrip antenna concept

There is no doubt that viable alternatives to metal waveguide slotted arrays can now be constructed using the slotted triplate or cavity backed techniques as listed above. Compared with waveguide antennas these new types have advantages in weight, thickness and cost but do involve a number of assembly steps in manufacture. The ideal arrangement would allow the conductors together with the feed to be printed onto a single substrate directly backed with the ground plane. Such an assembly would show a further saving in weight, assembly cost and allow the resulting thin structure to be mounted conformally onto the surface of a vehicle or missile etc. In essence, this takes the cavity-backed printed radiator to the extreme case where there is no air gap and the printed conductor and feed lies on the upper substrate surface while the lower substrate surface is backed with the ground plane. This leads directly to the microstrip antenna concept as defined at the beginning of this Chapter, and some examples showing the feeding arrangement are given in Fig. 1.13. The first published reference to a microstrip antenna array that we are aware of is a patent application by Gutton and Baissinot (1955) three years after the conception of the microstrip transmission line itself (Greig and Englemann, 1952). Apart from one paper on radiation effects, (Lewin, 1960) and patent action by Cashen *et al.* (1970), the main interest in the microstrip antenna concept did not emerge until after about 1970. It was probably the need for con-formal missile and spacecraft antennas that provided the impetus rather than the good selection of microstrip substrates that were becoming available.

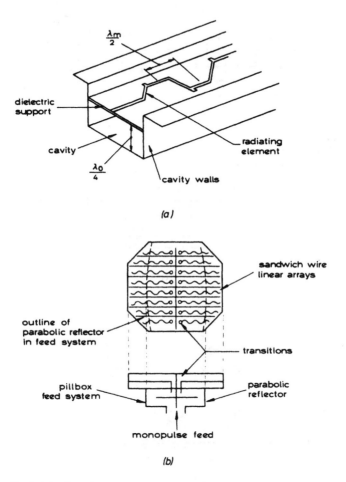

(a)

(b)

Fig. 1.11   *Sandwich wire antenna*
   a Basic configuration
   b Monopulse array showing pill-box feed in section

1.4.1 Fundamental limitations of microstrip antennas

It is possible to deduce the fundamental limitations of microstrip antennas, at least in qualitative outline, without examining specific designs in detail. Experience with slotted triplate and cavity backed printed antennas shows that the bandwidth decreases with the distance of separation between the radiating element and the ground plane, so thinner antennas have less bandwidth. Bandwidth in this sense is dictated by the frequency range over which the input match is acceptable but constraints on the radiation pattern performance may also dictate the antenna bandwidth. On this basis, one may expect the bandwidth of the very thin microstrip

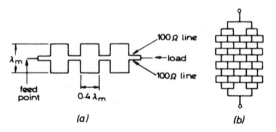

(a)   (b)

Fig. 1.12   *Chain antenna*
  a Basic linear element
  b Two-dimensional array

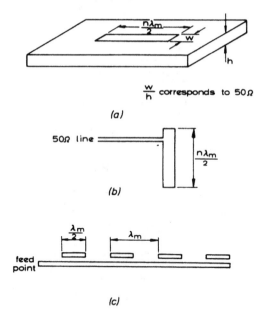

Fig. 1.13   *Microstrip radiating elements*
  a Narrow-line resonant radiating element
  b Direct-feed connection
  c Proximity fed resonant radiators

antennas to be further reduced and this is definitely the case. An explanation is provided by the well known supergain concepts (Chu, 1948) which related antenna size to its bandwidth. The microstrip antenna occupies less volume than the cavity backed antennas and hence less bandwidth can be expected.

This explanation is confirmed in later Chapters where it is shown that, whereas the radiation power of a microstrip antenna with constant applied voltage is essentially independent of the substrate height (h), the energy stored is inversely proportional to h. The antenna Q-factor is therefore also inversely proportional to h, since

$$Q_r = \frac{2\pi f_r \, \mathcal{E}_s(1/h)}{P_r}$$

where $\mathcal{E}_s(1/h)$ is the stored energy, P_r is the power radiated and f_r is the frequency. When conduction and dielectric losses are taken into account, the situation becomes more complicated because the power absorbed by these two mechanisms vary in different ways with the substrate height for constant applied voltage. The total Q is given by

$$Q_T = \frac{2\pi f_r \, \mathcal{E}_s(1/h)}{P_r + P_d + P_c}$$

showing that losses increase the bandwidth, which may appear to be an advantage. It must, however, be remembered that the antenna efficiency η depends upon the ratio of radiated power to total input power, so

$$\eta = \frac{P_r}{P_r + P_d + P_c} \times 100\%$$

and any increase in bandwidth due to losses is thus matched by a proportional reduction in efficiency. For typical materials, the radiated power dominates the resonant behaviour of the antenna unless the substrate is extremely thin, and in practice, Q-factors in the range 20–200 are commonly tolerated to gain the low profile advantage of the microstrip configuration.

The next problem concerns the fact that the feed structure is often printed on the substrate surface together with the radiating elements. The feeder lines introduce additional loss and a small amount of power can be coupled from one feeder to another by surface wave action in the dielectric substrate. This makes the antenna aperture distribution and hence sidelobe level difficult to control as does the complex higher mode action occurring in the triplate slotted array. The feeders on the microstrip substrate can, however, directly radiate and further contribute to the radiation pattern degradation, so some form of screened feeder is necessary when even moderate sidelobe levels are required unless some other technique can be found. Unlike a slot element, the microstrip radiating element is surround by a substrate that readily supports surface waves as the frequency increases and a small amount of power is injected into the substrate at each radiator. Scattering of these surface waves at the edge of the substrate board can further aggravate the control of sidelobes.

Finally, mechanical tolerances are likely to be a key factor in these thin microstrip assemblies imposing a limit on the precision with which the aperture phase and amplitude distribution can be controlled in manufacture. To this must be added the tolerance on the electrical parameters of the material and the mechanical characteristics in temperature, aging effects etc. of the substrate. The thicker cavity backed antennas with air gaps would appear to be less dependent on mechanical and electrical tolerances but require more intricate assembly in manufacture. In order to obtain a balanced view a system designer needs to weigh the overall advantages and disadvantages of each type of antenna and Table 1.1 summarises some of the main points made.

Table 1.1   *Comparison of flat-profile antenna types*

	Advantages	Disadvantages
Microstrip antennas	1. Thin profile 2. Light weight 3. Simple manufacture 4. Can be made conformal 5. Low cost 6. Integration with circuits	1. Extraneous radiation from feeds, junctions and surface waves 2. Low efficiency 3. Smaller bandwidth 4. Additional tolerance problems
Triplate	1. Thin profile 2. Light weight 3. Shielded feed lines 4. Can be made conformal	1. Complex manufacture 2. Narrow bandwidth 3. Coupling occurs inside structure 4. Tolerance problems
Cavity-backed printed antennas	1. Low line loss 2. Action generally well understood 3. Operation at medium power levels	1. Thick profile (especially at low frequencies) 2. Narrow bandwidth (except sandwich wire antenna) 3. Difficult to make conformal
Waveguide	1. Low line loss 2. Operation well understood 3. Operation at high power level	1. Thick profile (especially at low frequencies) 2. Complex manufacture 3. Difficult to make conformal

1.4.2 Performance trends

The above has indicated that the considerable reduction in thickness offered by the microstrip antenna concept can only be obtained at the expense of some degradation in performance parameters. It is possible to show this clearly by comparing typical performance figures for the various types of flat plate antennas. Hall and James (1978) have collated recent data, but the specifications aimed at are different for each antenna and a strict comparison is not possible. A pronounced trend, illustrating the degradation of efficiency, bandwidth and sidelobe level, can however be extracted as shown in Table 1.2, which is confined to types of travelling-wave antennas which are either air or dielectric filled. Table 1.2 is to be regarded as a first-order assessment of the state-of-the-art up to about 1979 based on our recent experience and literature collation. The trade-off between performance and antenna thickness is very evident for the dielectric filled antenna structures. The most significant property is, however, the low manufacturing cost associated with the microstrip

Table 1.2   *Approximate realisable performance of various types of travelling-wave antennas based on state-of-the-art information*

Type	Efficiency %	Bandwidth % of centre frequency	Feasible sidelobe level dB	Typical cost relative to slotted waveguide	Typical thickness of planar array (without radome) λ_0
Slotted-waveguide (air spaced)	79-90	10-15	30 35	1·0	0·28-0·5
Cavity-backed printed elements (sandwich wire/air spaced)	80 90	30-40	27 30	0·3 -0·7	0·25-0·4
Slotted triplate (dielectric filled)	50-80	2-6	22-26	0·1 -0·3	0·06-0·12
Microstrip	40-70	1-4	20-25	0·01-0·08	0·02-0·05

concept which together with the other advantages in Table 1.1 is the reason for the great interest that has developed. The main research aim is clearly to investigate ways of improving the efficiency, bandwidth and sidelobe level control without sacrificing the cost and operational advantages. New types of microstrip antenna array configurations will almost certainly be devised around the generic forms, i.e. resonant or travelling-wave type, corporate fed etc. The important question about the extent to which further improvements can be made can only be answered in time but it is encouraging to note that the newer flat-plate antenna techniques have received much less concentrated investigation than slotted waveguide arrays which themselves are still being improved.

1.5 Summary comment

The basic constructional techniques employed over the past three decades to create thinner, lighter, lower cost flat-plate antenna structures have been reviewed. In this context, the microstrip antenna concept takes its place as a logical means of creating a very thin planar antenna structure. It is evident that the antenna efficiency, bandwidth and sidelobe level control must in general be traded for the reduction in the thickness of the antenna assembly. For microstrip this is accompanied by a drastic drop in manufacturing costs, which, together with other operational advantages can make a microstrip antenna more attractive to the system designer than a conventional slotted waveguide having superior electrical performance. Antenna efficiency, bandwidth and sidelobe level control are therefore identified as the main areas for intensified research in microstrip antennas which is most likely to centre around well known generic forms common to other types of flat plate antennas. Bearing in mind that research effort has only recently been directed towards microstrip antennas their full potential performance capability is unlikely to have been reached.

1.6 References

ADAMS, A. T. (1967): 'Flush mounted rectangular cavity slot antennas: theory and design', *IEEE Trans.*, **AP-15**, pp. 342–51

BAZIRE, T. W., *et al.* (1975): 'A printed antenna radome (RADANT) for airborne doppler navigation radar'. Proc. conf. on Antennas for Aircraft and Spacecraft. IEE Conf. Publ. 128, pp. 35–40

BREITHAUPT, R. W. (1968): 'Conductance data for offset series slots in stripline', *IEEE Trans.*, **MTT-16**, pp. 969–70

CAMPBELL, J. G. (1969): 'An extremely thin omnidirectional microwave antenna array for spacecraft applications'. NASA Technical Note D-5539

CASHEN, E. R., *et al.* (1970): 'Improvements relating to aerial arrangements'. British Provisional Patent (EMI LTD) Specification No. 1294024

CHU, L. J. (1948): 'Physical limitations of omni-directional antennas', *J. Appl. Phys.*, **19**, pp. 1163–1175

CLAVIN, A., *et al.* (1974): 'An improved element for use in array antennas', *IEEE Trans.*, **AP-22**, pp. 521–6

COHN, S. B. (1954): 'Characteristic impedance of shielded-strip transmission line', *IRE Trans.*, **MTT-2**, pp. 52–55

COLLIER, M. (1977): 'Microstrip antenna array for 12 GHz TV', *Microwave J.*, **20**, (9), pp. 67–71

COX, R. M., and RUPP, W. E. (1970): 'A circularly polarised array element', *IEEE Trans.*, **AP-18**, pp. 804–07

CRAVEN, G. F., and HOCKHAM, G. A. (1975): 'Waveguide antennas'. British Patent Specification No. 1409749

CULLEN, A. L., and GOWARD, F. K. (1946): 'The design of waveguide fed arrays to give a specified radiation pattern', *J.I.E.E.*, **93**, Pt. IIIA, pp. 683–92

DUBOST, G., *et al.* (1976): 'Theory and applications of broadband microstrip antennas'. Proc. 6th European Microwave Conference, Rome, pp. 275–279

FRITZ, W. A., and MAYES, P. E. (1974): 'A frequency scanning stripline fed periodic slot array'. Proc. of the IEEE/AP-S International Symposium, Atlanta, Georgia, pp. 278–81

FUBINI, E. G., *et al.* (1955): 'Stripline radiators'. IRE Nat. Con. Rec., **3**, pp. 51–55

GALEJS, J. (1962): 'Excitation of slots in a conducting screen above a lossy dielectric half-space', *IRE Trans.*, **AP-10**, pp. 436–443

GALEJS, J. (1963): 'Admittance of a rectangular slot which is backed by a rectangular cavity', *IEEE Trans.*, **AP-11**, pp. 119–26

GRAHAM, R., *et al.* (1974): 'Monopulse aerials for airborne radars'. Proc. 4th European Microwave Conference, Montreux, pp. 362–366

GREEN, H. E., and WHITROW, J. L. (1971): 'Analysis of sandwich wire antennas', *IEEE Trans.*, **AP-19**, pp. 600–605

GREIG, D. D., and ENGLEMANN, H. F. (1952): 'Microstrip – a new transmission technique for the kilomegacycle range', *Proc. IRE*, **40**, pp. 1644–1650

GUTTON, H., and BAISSINOT, G. (1955): 'Flat aerial for ultra high frequencies'. French Patent No. 703113

HALL, P. S., and JAMES, J. R. (1978): 'Survey of design techniques for flat profile microwave antennas and arrays', *Rad & Electron. Eng.*, **48**, pp. 549–565

HALL, P. S., WOOD, C., and JAMES, J. R. (1981): 'Recent examples of conformal microstrip antenna arrays for aerospace applications'. 2nd I.E.E. Int. Conf. on Ant. and Prop., York

HARVEY, A. F. (1963a): 'Microwave engineering' (Academic Press, London) pp. 634–5

HARVEY, A. F. (1963b): 'Microwave engineering' (Academic Press, London) pp. 636–7

HERSCH, W. (1973): 'Very slim high gain printed circuit microwave antenna for airborne blind landing aid'. Proc. AGARD Conf. 139 Antennas for Avionics, Munich, pp. 9–1 to 9–2

HILL, R., and PAUL, D. H. (1975): 'A rapid versatile method of designing printed stripline aerial arrays'. Mullard Research Laboratories Annual Review, pp. 36–37

HWANG, T. L., RUTLEDGE, D. B., and SCHWARZ, S. E. (1979): 'Planar sandwich antennas for submillimetre applications', *Appl. Phys. Lett.*, 34, pp. 9–11

ITOH, T., and HEBERT, A. S. (1978): 'Simulation study of electronically scannable antennas and tunable filters integrated in a quasi-planar dielectric waveguide', *IEEE Trans.*, MTT-26, pp. 987–991

JAMES, J. R., and HENDERSON, A. (1978a): 'Electrically short monopole antennas with dielectric or ferrite coatings', *Proc. IEE*, 125, pp. 793–803

JAMES, J. R., and HENDERSON, A. (1978b): 'Investigation of electrically small VHF and HF cavity-type antennas'. Proc. Inter. Conf. on Antennas and Propagation. IEE Conf. Publ. 169, Part 1, pp. 322–326

JASIK, H. (1961): 'Antenna engineering handbook' (McGraw-Hill, New York) chap 9

JOSEFSON, L., *et al*. (1974): 'A stripline flat antenna with low sidelobes'. Proc of the IEEE/ AP-S International Symposium, Atlanta, Georgia, pp. 282–285

KEEN, K. M. (1974): 'A planar log-periodic antenna', *IEEE Trans.*, AP-22, pp. 489–490

KLOHN, K. L., *et al*. (1978): 'Silicon waveguide frequency scanning linear array antenna', *IEEE Trans.*, MTT-26, pp. 764–773

KNITTEL, G. H. (1972): 'Design of radiating elements for large, planar arrays: accomplishments and remaining challenges', *Microwave J.*, 15, pp. 27–28, 30, 32, 54, 80

LAGERLOF, R. O. E. (1973): 'Optimisation of cavities for slot antennas', *Microwave J.*, 16, pp. 12C–12F

LAURSEN, F. (1973): 'Design of periodically modulated triplate antenna'. Proc. AGARD Conf. 139, Antennas for Avionics, Munich, pp. 3–1 to 3–5

LEWIN, L. (1960): 'Radiation from discontinuities in stripline', *Proc. IEE*, 107C, pp. 163–170

McDONOUGH, J. A. (1957): 'Recent developments in the study of printed antennas'. IRE Nat. Con. Rec., 5, Pt. 1, pp. 173–176

OLINER, A. A. (1954): 'The radiation conductance of a series slot in strip transmission line'. IRE Nat. Con. Rec., 2, Pt. 8, pp. 89–90

OLINER, A. A., and MALECH, R. G. (1966a): 'Radiating elements and mutual coupling' *in* HANSEN, R. C. (Ed.): 'Microwave scanning antennas' (Academic Press, London) Vol. 2, p. 111

OLINER, A. A., and MALECH, R. G. (1966b): 'Radiating elements and mutual coupling' *in* HANSEN, R. C. (Ed.): 'Microwave scanning antennas' (Academic Press, London) Vol. 2, p. 137

OLINER, A. A. (1957): 'The impedance properties of narrow radiating slots in the broad face of rectangular waveguides, parts I and II', *IRE Trans.*, AP-5, p. 4

PALUMBO, B., and COSENTINO, S. (1973): 'Circularly polarised L-band planar array for aeronautical satellite use', Proc. AGARD Conference on Antennas for Avionics, Munich, Conf. Publ. 139, pp. 22–1 to 22–15

RAO, J. S., and DAS, B. N. (1978): 'Impedance of off-centred stripline fed series slot', *IEEE Trans.*, AP-26, pp. 893–895

RODGERS, A. (1972): 'Wideband squintless linear arrays', *Marconi Rev.*, 35, pp. 221–43

ROTMAN, W., and KARAS, N. (1957): 'The sandwich wire antenna: a new type of microwave line source radiator'. IRE Nat. Con. Rec, 5, pp. 166–172

SIDFORD, M. J. (1975): 'Performance of an L-band AEROSAT antenna system for aircraft.' Proc. Conference on Antennas for Aircraft and Spacecraft. IEE Conf. Publ. 128, pp. 123–129

SILVER, S. (1965a): 'Microwave antenna theory and design' (Dover Publications Inc., New York) pp. 321–7

SILVER, S. (1965b): 'Microwave antenna theory and design' (Dover Publications Inc., New York) p. 329

STARK, L. (1972): 'Comparison of array element types' *in* OLINER, A. A. and KNITTEL, G. H. (Eds.): 'Phased array antennas'. Array Antenna Symposium, (Artech House, Dedham, Mass., 1972) pp. 51–66

STEVENSON, A. F. (1948): 'Theory of slots in rectangular waveguides', *J. Appl. Phys.*, **19**, p. 24

TIURI, M., *et al.* (1976): 'Printed circuit radio link antenna'. Proc. 6th European Microwave Conf., Rome, pp. 280–282

WATSON, W. H. (1946): 'Resonant slots', *J.I.E.E.*, **93**, Pt. IIIA, pp. 747–77

WILKINSON, W. C. (1974): 'A class of printed circuit antenna'. Proc. of IEEE/AP-S International Symposium, Atlanta, Georgia, pp. 270–273

YOSHIMURA, Y. (1972): 'A microstripline slot antenna', *IEEE Trans.*, **MTT-20**, pp. 760–762

Chapter 2

Microstrip design equations and data

It is well known among antenna designers that the performance obtained from a slotted waveguide travelling-wave array is very dependent on the precision with which the waveguide itself is designed and manufactured. If a parallel feeding method is used the design is, of course, less critical, but then cost, size and weight factors are adversely affected. It is apparent from Chapter 1 that the accuracy to which the aperature distribution of all the flat-plate antennas can be designed is also dictated by the design of their respective transmission-line structures. The design of microstrip antennas therefore largely centres around the transmission line properties of microstrip lines themselves, but there is, however, a major problem with microstrip in as much that no exact design equations exist in simple closed mathematical form. For instance, the field behaviour within a rectangular metal waveguide can be rigorously expressed in terms of simple trigonometric formula; the more difficult waveguide problem of the effect of rough metal surfaces, deformed sides, the presence of holes and obstacles etc. can be analysed with a degree of accuracy sufficient for component design (Marcuvitz, 1951). By comparison, the wavelength λ_m of the dominant wave in a microstrip line can only approximately be represented by simple equations (Gunston, 1972) (Hammerstad and Bekkadal, 1975) (Hammerstad, 1975) and cut-and-try methods are generally relied upon to obtain precise dimensions. To obtain the field structure in the microstrip line and/or express λ_m with a precision demanded by most design specifications, it is necessary to compute equations involving extensive mathematical series but this in turn leaves some doubt about the resulting numerical accuracy and somewhat defeats the object from a precision design standpoint. The characterisation of discontinuities in microstrip lines such as steps, tapers, bends etc. can be assessed at low frequency using quasi-static techniques which embody the assumption that radiation effects can be completely neglected. Consequently the resulting data is of very limited use in microstrip antenna design giving at best some indication of how to shorten the line lengths to tune up radiating elements. Virtually no design data has previously existed on microstrip discontinuities in the higher frequency ranges of interest to antenna designers where radiation is, of course, a dominant effect. These high frequency aspects are the subject of Chapter 3. It should be noted that the literature refers to full-wave as opposed to quasistatic analyses implying that the high-frequency behaviour of microstrip discontinuities has in fact been addressed, but radiation effects are almost without exception omitted. A survey of these quasistatic and full-wave solutions for both microstrip and slot line discontinuities has recently been given by Gupta *et al.* (1979).

The situation is, therefore, that microstrip lines and discontinuities cannot be calculated as yet with sufficient precision to obviate the need for cut-and-try experimental methods and the dependence on measurements increases with frequency. To some extent, these comments are applicable to very long waveguide slot arrays which ultimately have to be experimentally trimmed but the point we make here is that microstrip design equations are less accurate than their waveguide counterparts to start with. The purpose of this Chapter is to collate existing design equations for microstrip lines and discontinuities and indicate their usefulness in microstrip antenna design. The simpler treatments are dealt with first and the chapter concludes with some consideration of the effects of manufacturing tolerances on the electrical parameters.

2.1 Equations based on TEM properties

The sectional sketch of a microstrip line, Fig. 2.1, shows the conductor width and thickness, the substrate height and relative permittivity w, t, h and ϵ_r, respectively.

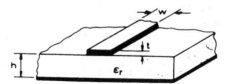

Fig. 2.1   *Sectional sketch of microstrip line*

Unless specified, the substrate relative permeability μ_r will be taken as unity, and in most cases of practical interest the finite strip thickness can be neglected. A reader requiring a detailed discussion of the physical action should consult one of the numerous treatments referenced in Gunston (1972) or Hammerstad and Bekkadal (1975). For design purposes, we require a knowledge of the wavelength λ_m of the wave guided in the microstrip and also the characteristic impedance Z_m of the line. The keystone of the calculation procedure rests on the fact that the structure would be readily analysed if the dielectric material occupied all space; the conducting strip together with its image in the ground plane is then capable of supporting a pure transverse electromagnetic (TEM) wave and the characteristic line impedance is

$$Z_d = \frac{1}{v_d C_d} \ \Omega \qquad\qquad (2.1)$$

where v_d = velocity of propagation along the line (m/s) and C_d = line capacity per unit length (F/m). When the dielectric material is completely removed, we denote the corresponding parameters of the air-spaced line by Z_0, v_0 and C_0 and these are related to the dielectric-filled line parameters by

$$v_d = v_0/\sqrt{\epsilon_r} \quad ; \quad C_d = C_0\epsilon_r$$

$$v_0 = 1/\sqrt{\mu_0\epsilon_0} \quad ; \quad Z_d = \frac{Z_0}{\sqrt{\epsilon_r}} \tag{2.2}$$

Now when the dielectric material only partially fills the space above the ground plane as in microstrip, it is physically reasonable to expect the corresponding values of velocity, capacity and impedance denoted by v_m, C_m and Z_m, respectively to lie somewhere between the values appropriate to the two extreme cases of a completely dielectric filled and empty space thus $v_d < v_m < v_0$, $C_0 < C_m < C_d$ and $Z_d < Z_m < Z_0$. This is indeed so, and it is convenient to introduce the concept of a filling factor q, $(0 \leqslant q \leqslant 1)$ and effective relative permittivity ϵ_e $(1 \leqslant \epsilon_e \leqslant \epsilon_r)$ where

$$\epsilon_e = 1 + q(\epsilon_r - 1), \qquad v_m = \frac{v_0}{\sqrt{\epsilon_e}}$$

$$C_m = C_0\epsilon_e, \qquad Z_m = \frac{Z_0}{\sqrt{\epsilon_e}}$$

$$\lambda_m = \frac{\lambda_0}{\sqrt{\epsilon_e}}, \qquad \beta = \frac{2\pi}{\lambda_m} \tag{2.3}$$

λ_0 is the wavelength in the air-filled line and is identical to the wavelength in free space. Now C_0 and hence Z_0 $(= 1/v_0 C_0)$ can be readily calculated by a conformal transformation method; Assadourian and Rimai (1952) carried out the earlier work on this which has subsequently been refined and extended by Wheeler (1964). The modification of this analysis to allow q to be calculated then followed by Wheeler (1965) who obtained straightforward expressions for Z_m and λ_m. Wheeler has subsequently published numerous papers on this technique and Wheeler (1977) contains an interesting overview of his important contributions to this topic area.

It would appear from eqn. 2.3 that the microstrip line is completely solved once q is known, but this is not so because the structure cannot support the TEM mode but rather a set of discrete hybrid modes having nonzero E_z and H_z components. It is found, however, that the lowest-order mode strongly resembles a TEM wave at low frequency. It has become the practice to refer to this wave as a quasi-TEM wave and use eqn. 2.3 with q derived by conformal transformation, to represent the low-frequency behaviour of the line. Unfortunately, the expressions do not embrace all values of w/h and much effort has been expended investigating how best to subdivide the range of w/h values to obtain the smoothest and most accurate transition. For instance, Wheeler's basic result for Z_m (or Z_0 for $\epsilon_r = 1$) is

$$Z_m = \frac{377}{\pi\sqrt{2}\sqrt{\epsilon_r + 1}} \left\{ \ln\left(\frac{8h}{w}\right) + \frac{1}{32}\left(\frac{w}{h}\right)^2 \right.$$

$$\left. - \frac{1}{2}\left(\frac{\epsilon_r - 1}{\epsilon_r + 1}\right)\left(\ln\frac{\pi}{2} + \frac{1}{\epsilon_r}\ln\frac{4}{\pi}\right)\right\} \; \Omega$$

for $w/h < 1$ (2.4a)

$$Z_m = \frac{377}{2\sqrt{\epsilon_r}} \left\{ \frac{w}{2h} + 0.441 + 0.082\left(\frac{\epsilon_r - 1}{\epsilon_r^2}\right) \right.$$

$$\left. + \left(\frac{\epsilon_r + 1}{2\pi\epsilon_r}\right)\left\{1.451 + \ln\left(\frac{w}{2h} + 0.94\right)\right\}\right\}^{-1} \Omega$$

for $w/h > 1$ (2.4b)

Alternatively, a formula due to Schneider (1969a) is

$$\epsilon_e = \frac{1}{2}\left\{\epsilon_r + 1 + (\epsilon_r - 1)\left(1 + \frac{10h}{w}\right)^{-1/2}\right\}$$ (2.5)

which may be used in conjunction with eqn. 2.4 for the case $\epsilon_r = 1$, and eqn. 2.3. These formulas, either eqn. 2.4 alone or eqn. 2.5 together with eqn. 2.4 enable design curves to be derived for the parameters of eqn. 2.3. Some correction for the strip thickness t can be allowed for by substituting an effective width w' for w where

$$w' = w + \frac{t}{\pi}\left(1 + \ln\left(\frac{2x}{t}\right)\right)$$ (2.6)

with $x = h$ for $w > (h/2\pi) > 2t$ and $x = 2\pi w$ for $(h/2\pi) > w > 2t$. Hammerstad and Bekkadal (1975) have suggested refinements to these expressions for ϵ_e and Z_m to obtain increased accuracy, but it is not entirely clear what low-frequency range limitations must be imposed. Clearly, these static field calculations become progressively more inaccurate as the frequency is raised. Analysis by Getsinger (1973) shows that the dispersion effects can be corrected for to some extent by calculating

$$\epsilon_e(f) = \epsilon_r - \frac{\epsilon_r - \epsilon_e}{1 + F_1\{f/f_p\}^2}$$

$$f_p = \frac{0.4Z_m}{h[\text{mm}]} \; [\text{GHz}], \qquad F_1 \sim 0.6 + 0.009\,Z_m$$ (2.7)

where $\epsilon_e(f)$ and ϵ_e are the values of the effective relative permittivity at a frequency f GHz and at zero frequency, respectively. Now, for microstrip antenna design purposes, dispersion effects will be generally significant and if omitted could lead to errors in excess of those generated by the static formulas of eqns. 2.4 and 2.5 at crossover values in the w/h range. We therefore regard the above static field equations together with Getsinger's dispersion correction as first-order equations for antenna design purposes such as calculating the phase relationship along a micro-strip feeder or stub element. A computer programme (IPED) based on eqns. 2.4 and 2.7 with some modifications to the latter following recommendations by Owens (1976), has been used by the present authors and a sample print-out of Z_m and ϵ_e together with dispersion corrections for ϵ_e are given in Appendix A. Readers wish-ing to make their own modifications to these static field equations will find copious details in the references contained in Gunston (1972), Hammerstad and Bekkadal (1975) and Gupta *et al.* (1979).

2.2 Estimation of losses in microstrip

The dissipative losses associated with microstrip lines are one of the major limi-tations with microstrip antennas. As we have discussed at the commencement of this Chapter, the radiating regions of a microstrip antenna array, whether they are discrete or continuous elements, are fed with power by stripline feeders and it is within the latter that the major dissipative loss takes place. It is not untypical for half the available power to be dissipated in the feeders, and for many applications this is unacceptably high. The salient problem is to find new ways of constructing microstrip antennas to reduce this loss while retaining the geometrical simplicity. By comparison, the problem of being able to accurately predict this loss is of less design significance since the loss mechanisms only have a second-order effect on the microstrip wavelength and almost no bearing on the radiation mechanism. It may be sufficient in some antenna array designs to establish whether the overall loss due to the microstrip feeders is, say, around 2dB, or near to, say, 4dB, in which case a very elementary loss estimation will suffice. The situation is therefore that although research into improved loss analysis continues to be of interest, the existing loss equations are quite adequate for microstrip antenna design, at least with the present antenna requirements. We will now list the loss formulas that will be found useful and the effects associated with the dielectric and metal strip can be treated individually.

2.2.1 Dielectric loss
Typical dielectric substrate material creates a very small power loss at microwave frequencies. The calculation of dielectric loss in a dielectric filled transmission line is easily carried out provided exact expressions for the wave mechanism are avail-able but for microstrip this involves extensive mathematical series and numerical methods. A coarse approximation is to assume that the substrate material, having a complex relative permittivity $\epsilon_r - j\,\epsilon_r'$, exists uniformly over all space in which case the attenuation constant is denoted by α_{du}:

$$\alpha_{du} = \frac{\omega}{2}(\mu_0 \epsilon_0 \epsilon_r)^{1/2} \tan\delta \qquad (2.8)$$

where $\tan\delta = \epsilon_r'/\epsilon_r$; this also assumes TEM wave transmission. To improve the accuracy of eqn. 2.8, it seems reasonable to invoke the effective relative permittivity

ϵ_e together with an effective loss tangent. The latter, however, can again be expressed in terms of ϵ_e (Welch and Pratt, 1966), (Schneider, 1969b) to yield the attenuation constant α_d of a microstrip line, in the form

$$\alpha_d = 27.3 \, \frac{\epsilon_r}{\epsilon_e} \, \frac{(\epsilon_e - 1)}{(\epsilon_r - 1)} \, \frac{\tan \delta}{\lambda_m} \text{ dB/cm} \tag{2.9}$$

or alternatively,

$$\alpha_d = 4.34 \, \frac{1}{\sqrt{\epsilon_e}} \left(\frac{\epsilon_e - 1}{\epsilon_r - 1} \right) \sqrt{\frac{\mu_0}{\epsilon_0}} \, \sigma_d \text{ dB/cm} \tag{2.10}$$

where σ_d is the conductivity of the dielectric material and unit of length is cm.

These formulas are quite adequate for the majority of microstrip antenna applications at microwave frequencies using low loss dielectric material ($\epsilon_r \sim 2$ to 10) substrates. Eqn. 2.10 is more suitable for lossier substrates. It seems unlikely that lossier substrates would be used in antenna applications but at millimeter wavelengths the dielectric loss could be more significant.

2.2.2 Conductor loss

This is by far the more significant loss effect over a wide frequency range and is created by the high current density in the edge regions of the thin conducting strip. Surface roughness and strip thickness t also have some bearing on the loss mechanism. The simplest estimate of the attenuation constant attributable to conductor loss alone is

$$\alpha_c \sim \frac{8.68 \, R_s}{w \, Z_m} \text{ dB/cm} \; ; \qquad R_s = \left(\frac{\omega \mu_0}{2 \sigma_c} \right)^{1/2} \tag{2.11}$$

where R_s is the surface resistivity (Ω per square) of both sides of the strip and $\sigma_c = $ conductivity of strip. Eqn. 2.11 apparently gives excessive values of α_c and Pucel *et al.* (1968) have presented much improved formulas for α_c which are reasonably substantiated by measurements. Owing to the complexity of computing the current flow on the strip, they have utilised the 'incremental inductance rule' of Wheeler (1942). The method embraces the microstrip line impedance Z_m and hence various ranges of w/h must be considered as follows:

$w/h \leqslant \tfrac{1}{2}\pi$:

$$\alpha_c = \frac{8.68 \, R_s}{2\pi \, Z_m h} \left[1 - \left(\frac{w'}{4h} \right)^2 \right] \left\{ 1 + \frac{h}{w'} + \frac{h}{\pi w'} \left[\ln \left(\frac{4\pi w}{t} + 1 \right) \right. \right.$$

$$\left. \left. - \left(\frac{1 - t/w}{1 + t/(4\pi w)} \right) \right] \right\}$$

$\dfrac{1}{2\pi} < w/h \leqslant 2$:

$$\alpha_c = \frac{8.68 R_s}{2\pi Z_m h} \left[1 - \left(\frac{w'}{4h} \right)^2 \right] \left\{ 1 + \frac{h}{w'} + \frac{h}{\pi w'} \left[\ln \left(\frac{2h}{t} + 1 \right) \right.\right.$$

$$\left.\left. - \left(\frac{1 + t/h}{1 + t/(2h)} \right) \right] \right\}$$

$2 \leqslant w/h$:

$$\alpha_c = \frac{8.68 R_s}{Z_m h \left\{ \frac{w'}{h} + \frac{2}{\pi} \ln \left[2\pi e \left(\frac{w'}{2h} + 0.94 \right) \right] \right\}^2} \left[\frac{w'}{h} + \frac{w'/(\pi h)}{\frac{w'}{2h} + 0.94} \right]$$

$$\cdot \left\{ 1 + \frac{h}{w'} + \frac{h}{\pi w'} \left[\ln \left(\frac{2h}{t} + 1 \right) - \frac{1 + t/h}{1 + t/(2h)} \right] \right\} \tag{2.12}$$

where

$$w' = w + \frac{t}{\pi} \ln \left(\frac{4\pi w}{t} + 1 \right) ; \quad \frac{2t}{h} < w/h \leqslant \frac{1}{2\pi}$$

$$w' = w + \frac{t}{\pi} \ln \left(\frac{2h}{t} + 1 \right) ; \quad w/h \geqslant \frac{1}{2\pi}$$

where the definition of w' differs slightly from Gunston (1972) in eqn. 2.6; α_c is in dB/cm. Other loss equations have been derived by Schneider (1969b) and more recently Mirshekar-Syahkal and Davies (1979) have reported very detailed computations of losses using Legendre polynomial basis functions to represent the strip current. Computations of both α_c and α_d due to Bahl and Trivedi (1977) and collated by Gupta *et al.* (1979) are reproduced in Fig. 2.2 and illustrate the relative effects and frequency dependence for various substrates. Precise measurements of these loss effects is made difficult by other unavoidable loss mechanisms such as the radiation loss at coaxial to microstrip transitions and that due to material inhomogeneity. Measurements on the effects of surface roughness have been made by Van Heuven (1974) and agree well with theoretical predictions of Morgan (1949). The attenuation constant α_{cr} allowing for roughness effects can be doubled in some cases and Hammerstad and Bekkadal (1975) have proposed the empirical formula

$$\alpha_{cr} = \alpha_c \left\{ 1 + \frac{2}{\pi} \arctan 1.4 \left(\frac{\Delta}{\delta} \right)^2 \right\} \tag{2.13}$$

where Δ is the RMS surface roughness and the skin depth $\delta = 1/(R_s \sigma_c)$. The introduction of skin depth suggests that other effects may be observed if t is very small and calculations and measurements by Horton *et al.* (1971) indicate that while a very thin strip leads to excessive α_c values, there appears to be an optimum value of t corresponding to a strip thickness of only about three skin depths. This would appear to warrant further investigation but may introduce problems in manufacture regarding control of the printed conductor.

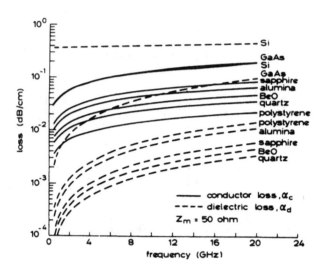

Fig. 2.2   *Conductor and dielectric losses as functions of frequency for microstrip lines on various substrates*
(Reproduced from Gupta *et al.* (1979))
$h = 0.635$ mm except for Ga As and Si where $h = 0.254$ mm

From the microstrip antenna design standpoint, surface roughness and conductor thickness are clearly factors that need to be considered, but their effects are likely to be small compared with the total conductor loss. The important question is whether some degree of geometrical freedom remains which would allow a reduction in α_c to be made without disturbing the manufacturing simplicity of the line. Increasing the thickness and rounding off the strip edges reduces the current density in the latter and hence α_c and comparative results are illustrated by Wheeler (1977). There appears to be little experimental evidence however on the dependence of α_c on conductor shape and the accuracy of all these loss formulas at higher frequencies is in any case questionable. For one thing transverse currents exist on the strip and some calculations on the strip current at high frequency have been given by Wiesbeck (1972). Since conductor loss is a major problem with large microstrip antenna arrays there would therefore appear to be some scope for altering the conducting strip geometry without creating an assembly which is complicated from both the design and manufacturing standpoint.

2.3 Higher-order analytical methods

2.3.1 Practical design implications

The analysis of microstrip has been a challenge to theorists for some two decades and a vast literature now exists. Many of the more recent methods are reviewed by Gupta *et al.* (1979). These theoretical investigations have greatly contributed to the understanding of the precise nature of the waves that can exist, but from a practical design standpoint the advance has been less significant. The simplified equations of Section 2.1 continue to form the basis of most circuit design procedures giving β to an accuracy within a few percent of the correct value and nominal values for Z_m; the line lengths are then trimmed by trial-and-error methods to obtain the final dimensions for manufacture. There are several aspects of this situation that are seldom emphasised enough and the main points are as follows:

(i) The exact mathematical formulations of microstrip culminate without exception in extensive numerical procedures involving truncated series, approximate integrals, matrix inversion etc. Numerical round-off errors and doubts about the precise convergence of the computations, although no more severe than in other numerical mathematical methods, can create an error bound that is of the order of the correction being sought to the quasi-TEM results. The same order of difference often exists between two exact formulations although comparisons are seldom made.

(ii) Experimental confirmation of theoretical β values invoke measurement errors which can be comparable to the small differences in the value under examination. See for instance the error bounds on measured data (Gupta *et al.*, 1978). In particular, it is not an easy matter to measure *in situ* the substrate permittivity at high frequency on a given length of microstrip.

(iii) The complete analysis of microstrip establishes that transverse currents flow on the conducting strip and several definitions of line impedance, some being functions of transverse coordinates, exist, (Schmitt and Sarges, 1971), (Krage and Haddad, 1972) (Bianco *et al.*, 1976) (Getsinger, 1979) (Arndt and Paul, 1979). The absence of a unique impedance concept means that the theoretical curves relating impedance and dispersion effects are of limited practical value, only giving some qualitative indication of matching conditions. It is now generally accepted that in comparison to β, the line impedance level is relatively less sensitive to dispersion effects and the impedance levels increase slightly with frequency.

(iv) The dependence of circuit design on trial-and-error adjustment of line lengths has prompted research into computer-aided iterative methods (Hosseini and Shurmer, 1978).

(v) With microstrip antennas, the microstrip feeders can be many wavelengths long. If large aperture phase errors are to be avoided line lengths must have the correct dimension to very close limits.

(vi) Microstrip circuits become good radiators at high frequency. The explanation of this is left to Chapter 3. It is important to appreciate therefore that microstrip antennas generally operate at frequencies where dispersion effects cannot be neglected. This is another reason for the need to trim line lengths experimentally in the absence of exact dispersion corrected values.

2.3.2 Methods of analysis

The dependence of design methods on the trial-and-error adjustment of line lengths will hopefully be removed by continued research into new analytical methods or the substantiation of an existing method as a reliable means of calculating β to the required accuracy. To this end it is useful to collate some of the more recent theoretical formulations and comment on their relative merits as far as can be ascertained.

As stated in item (vi) Section 2.3.1, antenna applications require analyses that deal with dispersion effects and as such solutions of Laplace's equation using conformal transformations, the finite difference method, the integral equation method and variational method in the Fourier transform domain do not embrace dispersion effects and will be omitted; these are reviewed in Gupta *et al.* (1979); Kaden (1974) presents additional refinements on the conformal transformation method.

One of the significant differences between the analysis of microstrip and conventional closed metal waveguides is that the fields in the latter can be represented by a sum of discrete eigenfunctions directly corresponding to the various waveguide modes, whereas microstrip requires a continuous spectrum of eigenfunctions. If a conducting shield is placed over the microstrip at some distance, it is possible to treat the structure as a closed waveguide using discrete eigenfunctions. It is convenient to separate these theoretical contributions accordingly into open microstrip (continuous spectrums) and shielded microstrip (discrete spectrums) and we will deal with the former first.

Probably the earliest attempt at analysing open microstrip was by Wu (1957), but his choice of solution has been questioned by Delogne (1970). Subsequently, Denlinger (1971) and Schmitt and Sarges (1971) presented similar contributions based on a Fourier integral representation of the transverse electric (TE) and transverse magnetic (TM) potentials. The main problem was that the results were critically dependent on the assumed current flow on the conducting strip. For narrow strips where $w \ll \lambda_0$ the transverse current (x-directed) can be neglected and the axial (z-directed) current $j_z(x)$ is assumed to be of the form

$$j_z(x) \sim (w^2 - 4x^2)^{-1/2}; \qquad |x| \leqslant \frac{w}{2} \tag{2.14}$$

by Schmitt and Sarges (1971) and

$$j_z(x) \sim 1 + \left| \frac{2x}{w} \right|^3 ; \qquad |x| \leqslant \frac{w}{2} \tag{2.15}$$

by Denlinger, (1971), in which case the coupled integral equations reduce to a single eigenvalue integral equation which is readily computed. Agreement with measurements is found to be very good for the dominant EH-mode. Van de Capelle and Luypaert (1973) have presented alternative forms for eqns. 2.14 and 2.15 to enable higher EH-modes to be calculated. None of these analyses deal with thick conductor strips with different current distributions existing on either side. Wiesbeck (1972) has derived current distributions on both sides by correcting static forms for displacement current. He also calculates transverse currents but the effect of a finite strip thickness on β is not investigated. The difficulty of finding the appropriate strip current distribution is avoided by the spectral domain method Itoh and Mittra (1973) whereby the boundary conditions are applied in the spectral domain rather than in the space domain and then the unknown current distributions are expanded in terms of known basis functions; both transverse and axial currents can be accommodated and the method is considered to be computationally simpler. Knorr and Tufekcioglu (1975) have used the same method to investigate the characteristic impedance variation with frequency using various impedance definitions; their results indicate changes of more than 10% with increasing frequency as do the results of Schmitt and Sarges (1971). The physical action that is responsible for dispersion has been the subject of other analytical methods (Van der Pauw, 1976) and (Kowalski and Pregla, 1972), while Chang and Kuester (1979) have included the effect of raising the conducting strip above the substrate; the latter paper is particularly valuable as regards revealing the wave mechanism.

Shielded microstrip was not considered until recent years (Brackelmann, 1967) and the basic idea is to surround the microstrip in a conducting box so that the structure becomes a loaded waveguide. The field expansions are a discrete series of modes which are truncated; the unknown amplitude coefficients can then be solved by a variety of numerical methods. Zysman and Varon (1969) obtain a pair of coupled homogeneous Fredholm integral equations of the first kind which are then solved by conversion into a matrix equation.

Hornsby and Gopinath (1969a) employ a Fourier analysis method which results in a set of homogeneous linear equations. A similar method which also embraces expansions for the current distributions is presented by Krage and Haddad (1972). Improved computational accuracy is claimed by Mittra and Itoh (1971) by invoking a singular integral equation approach and subsequently Itoh and Mittra (1974) have used the spectral domain technique in discrete form which, as in its application to open microstrip, requires a set of basis functions to represent the strip current. A completely different technique is based on the finite difference method of analysis (Hornsby and Gopinath, 1969b). This has been further developed by Corr and Davies (1972) to include higher-order modes.

The above short review of analytical methods indicates the great interest aroused in microstrip analysis and the better understanding of the wave mechanism that has resulted. It is unfortunate for the designer that little has been established regarding which method is most accurate and what error can be expected in the calculated β value. In all the methods, at least one series is truncated and the choice of the number of truncated terms must be decided upon by inspection of the convergence which may be obscured by round-off error. For open microstrip, the spectral domain method appears to be one of the most elegant techniques requiring fewer assumptions about the strip current behaviour for both open and shielded microstrip. The problem is then to ascertain the appropriate basis functions. The fact that shielded microstrip leads to a discrete, rather than an integral expansion, could be considered a computational advantage but this has not been established. Any such advantage could be negated by the need to consider the effect of the shield on the β values obtained; a further scrutiny of the convergence of β as a function of shield proximity is thus necessary.

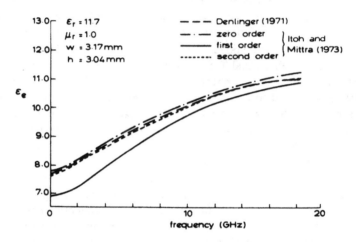

Fig. 2.3   *Comparative numerical results for open microstrip*
(Reproduced from Itoh and Mittra (1973)

Some comparative results are shown in Figs. 2.3 and 2.4 to illustrate the order of difference that can be obtained from some of the analyses. In Fig. 2.3, the dependence of the spectral domain method on the order of basis function and the agreement with Denlinger (1971) who takes an assumed current, is shown. In Fig. 2.4, some comparative results for shielded microstrip are given. More recently, Kuester and Chang (1979) have derived a canonical equation from which many of the other analyses can be shown to originate for the case when h and w are comparable. This important development enables a precise comparison to be made between a large number of methods and the results are reproduced in Figs. 2.5 and 2.6 showing a considerable discrepancy between the various data for ϵ_e and hence β. The problem is that the various analysis all use different forms for the strip current distribution and it is deduced by Kuester and Chang (1979) that the analyses of Jansen (1978), Fujiki *et al.* (1972), Kowalski and Pregla (1972), Pregla and Kowalski (1974), Van de Capelle and Luypaert (1973) and (1974), are likely to give the most accurate parameter values; the accuracy could be of the order $\pm 1\%$, which appears to be good until we consider the application to a long antenna feeder. For instance, an uncertainty in β of $\pm 1\%$ would create an aperture phase uncertainty of $\pm 36°$ at the end of a ten-wavelength long antenna series feeder line. Such a phase error is unacceptable for some antenna array specifications and endorses our assertion at the commencement of this Chapter about the need, at present, to trim long antenna feed lengths by trial and error methods.

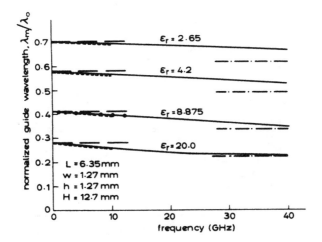

Fig. 2.4   *Comparative numerical results for shielded microstrip*
(Reproduced from Mittra and Itoh (1971).
H is the distance between the shield and ground plane and *L* the width of the enclosure
—— Mittra and Itoh (1971)
... open strip experiment
--- Zysman and Varon (1969)
—·— $1/\sqrt{\epsilon_r}$
— — Quasi TEM

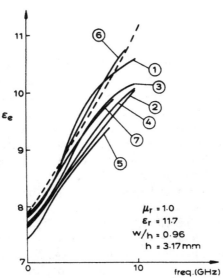

Fig. 2.5   *Comparison of effective relative permittivity computed by various authors*
(Reproduced from Kuester and Chang (1979)).
(1) Farrar and Adams (1972) (2) Itoh and Mittra (1973) (3) Van de Capelle and
Luypaert (1973) (1974) (4) Denlinger (1971) (5) Schmitt and Sarges ($\epsilon_r = 11.2$)
(1971) (6) Chang and Kuester (1979) (7) Pregla and Kowalski (1974)
— — — computed by programme IPED Appendix A

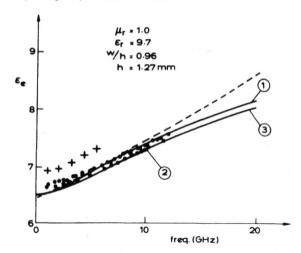

Fig. 2.6   *Comparison of effective relative permittivity computed by various authors*
(Reproduced from Kuester and Chang (1979))
(1) Schmitt and Sarges (1971) (2) Fujiki *et al.* (1972) (3) Kowalski and Pregla (1972)
w/h = 1·0
– – – computed by programme IPED Appendix A
Experimental results:
... Hartwig *et al.* (1968)
+ + + Deutsch and Jung (1970) $\epsilon_r = 9·8$, *w/h* = 1·0

2.4 Discontinuities in microstrip lines

Discontinuities in microstrip lines such as those depicted in Fig. 2.7 are commonly
an integral part of microstrip antennas, occurring both in the feeder lines and
radiating elements. Radiating elements can be regarded as desirable discontinuities
in the sense that the radiation loss created by the latter is usefully employed in the
antenna design whereas discontinuities in feeder lines create unwanted radiation.
Examples of the latter are bends, splitters, impedance steps etc., and their radiation
loss can corrupt the radiation pattern of the antenna. In extreme cases the feeders
need to be screened or converted to a triplate assembly as described in Chapter 6.
Even when the radiating loss can be neglected the impedance step or splitter in the
feeder must be characterised to control the design.

It was noted in item (vi) Section 2.3.1 that efficient microstrip antennas need to
be operated in a frequency range where substrate dispersion effects are not insignifi-
cant and the latter must be allowed for in the design of microstrip lines. By implication, the same requirement holds for the analysis of discontinuities. Unfortunately
to date, no analytical methods have been established that enable the dispersive and
radiation characteristics of microstrip discontinuities to be modelled with sufficient
accuracy. Thus the quasistatic and other methods can only be regarded as a means
of providing a rough estimation of the parameters of the discontinuity for antenna

design purposes. To this end, a note of some of the existing treatments of microstrip discontinuities is now given. It must be emphasised however that these analytical methods give quite useful data for conventional microstrip circuit design where radiation and dispersion effects are in general negligible.

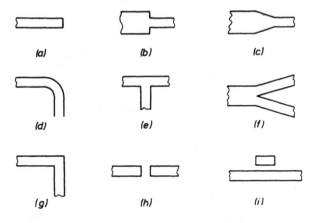

Fig. 2.7 *Commonly occurring microstrip elements containing discontinuities*
(a) Open-circuit termination *(b)* Step *(c)* taper *(d)* bend *(e)* T junction *(f)* Y junction *(g)* right-angle bend *(h)* gap *(i)* coupled resonator

2.4.1 Quasistatic analysis of discontinuities
In the quasistatic approach, the aim is to calculate the individual elements of the equivalent circuit of the discontinuity. There are problems involving the selection of a useful form for the equivalent circuit and the subsequent identification of each inductance and capacitance with the physical action which then has to be analysed. The principal methods employed are reviewed by Gupta *et al.* (1979) and include the treatments of Farrar and Adams (1972), Itoh *et al.* (1972), Silvester and Benedek (1973) and Thomson and Gopinath (1975). The computational nature of these treatments means that the results are focussed on specific cases and little or no design data is available.

2.4.2 Equivalent waveguide analysis of discontinuities
There are several ways of improving the quasistatic analyses to give some information about dispersion effects at higher frequency but the method which has recently been actively pursued is based on the equivalent waveguide model. The microstrip line of width w is considered to have an equivalent waveguide model width w_e which is derived from empirical data or represented by

$$w_e = \frac{120\pi h}{Z_m(\epsilon_e)^{1/2}} \tag{2.16}$$

This equivalent waveguide is bounded by both conducting walls and magnetic walls as shown in Fig. 2.8 and has the same phase constant β and impedance Z_m as the microstrip line. The idea is to convert the discontinuous microstrip configuration into its waveguide equivalent and apply a modal analysis such as for instance that described by Kuhn (1973). Some of the principal contributions on this method have been given by Wolff, Kompa and Mehran (1972), Mehran (1976), Menzel and Wolff (1977), Kompa (1978) and are reviewed in Gupta *et al.* (1979) who classify

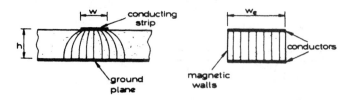

Fig. 2.8   *Sectional view of microstrip line and its equivalent waveguide model*

them as full-wave analyses. These methods do not include radiation loss at the discontinuity and it seems reasonable to assume that the resulting data can only account for dispersive effects at the lower frequency ranges; that is before the radiation loss becomes significant. Easter (1975) has reported experimental results on several types of discontinuities over the lower frequency ranges.

2.4.3 First-order closed-form equations for antenna design

It has already been concluded that a microstrip antenna designer will need to carry out measurements on microstrip discontinuities to determine their parameters, and subsequently correct for line lengths and widths. The above approximate treatments can be used as first-order design data but are not generally in suitable form for practical purposes. It is unlikely that designers will wish to continually engage in numerical techniques involving truncated series, matrix inversion etc., just to arrive at approximate parameters. The need for closed form equations that give a reasonable fit to the available theoretical results is thus apparent. The need for closed-form equations has been emphasied by Hammerstad and Bekkadal (1975) and Garg *et al.* (1978) who have curve fitted equations to available data on various discontinuties. These closed-form expressions for the commonly occuring microstrip discontinuities Fig. 2.9 are listed in Appendix B.

2.5 Manufacturing and operational aspects of microstrip substrates

There are many types of materials that are available for use as microstrip substrates but it is not easy to establish the tolerances on their electrical and mechanical parameters and their operational behaviour under shock, vibration, temperature cycling, moisture and aging. Harlan Howe (1974) gives useful data and states two obvious

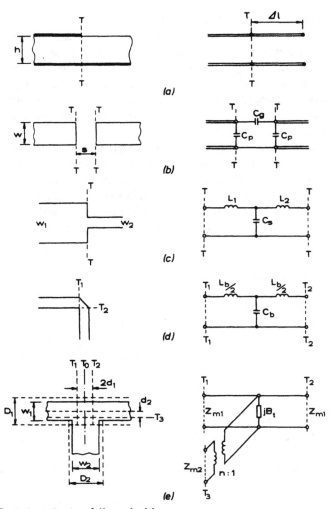

Fig. 2.9   *Equivalent circuits of discontinuities*
(a) Open circuit (b) gap (c) step (d) right-angle bend (e) T junction

difficulties. First, the lamination process is difficult to control for large sheets and significant variations in the parameters are to be expected in commercially available substrates. Second, the difficulty and cost of measuring these variations means that little is stated about the latter in product specifications. It must be appreciated that microstrip antennas are a relatively new application area for microstrip substrates and until recently no substrates had been manufactured specifically for antenna applications (Traut, 1979). The view could be taken, however, that antenna designers should try to make do with the available substrates in order to obtain a very low cost antenna but this depends on the antenna specification and is discussed in Chapter 8. Some comparative mechanical and electrical data for some commercially available substrates are given in Appendix C and some data on specific antenna configurations are included in later Chapters. It is noted that some of the substrates contain glass fibres which create anisotropic electrical characteristics of a somewhat random nature. This is another factor to take into consideration in antenna applications requiring precise radiation pattern control but the anisotropy can probably be neglected in less demanding situations. This is the only way in which anisotropic materials have so far been encountered in microstrip antenna design but the employment of well defined anisotropic substrate material such as sapphire cannot be ruled out in the future.

Little data is available about the power handling capability of microstrip at higher frequencies and this is no doubt mainly due to the relatively high conductor loss associated with microstrip. There may be antenna applications where sufficient power is available to make up for these losses in which case the rise in temperature of the antenna and voltage break down are the constraining criteria. Power handling considerations at lower frequencies are summarised in Gupta *et al.* (1979).

2.6 Summary comments

The literature on the theory of microstrip lines and their discontinuities is vast but almost without exception the methods do not embrace radiation effects and only a few deal with dispersion. This Chapter indicates which equations are useful to the microstrip antenna designer and assesses as far as possible the accuracy of parameters calculated.

It is concluded that even if exact mathematical formulations are used to compute β, the length of a long antenna feeder line will still require some experimental trimming. It is therefore recommended that the less exact but more convenient closed-form equations such as eqns. 2.1-2.7 be used to set up the lines prior to final experimental adjustment.

Microstrip discontinuities are commonly an integral part of antenna configurations but the theoretical methods for dealing with discontinuities are not yet developed enough to embrace dispersion and radiation effects. Experimental design techniques must be heavily relied upon but some first order empirically based equations (Appendix B) are recommended as a means of initiating the first iteration in an experimental design process.

The calculation of microstrip conduction and dielectric loss effects is not a critical factor in the deployment of microstrip radiation elements to create desired pattern characteristics. The overall antenna power loss is however a vital parameter and can limit the range of application of a particular type of antenna. The assessment of this loss is thus important and useful calculations can be carried out using the closed-form equations 2.9-2.13.

Finally, it is emphasised that tolerances in manufacturing processes and operational effects can impose limitations in some critical applications. A major problem is however acquiring sufficient data on all these effects. The main issues are itemised and an example of the typical data on commercially available substrates is given in Appendix C.

2.7 References

ARNDT, F., and PAUL, G. U. (1979): 'The reflection definition of the characteristic impedance of microstrip', *IEEE Trans*, **MTT-27**, pp. 724–731

ASSADOURIAN, F., and RIMAI, E. (1952): 'Simplified theory of microstrip transmission systems', *Proc. IRE*, **40**, pp. 1651–1657

BAHL, I. J., and TRIVEDI, D. K. (1977): 'A designer's guide to microstrip line', *Microwaves*, **16**, pp. 174–182

BIANCO, B., PARODI, M., and RIDELLA, S. (1976): 'On the definitions of characteristic impedance of uniform microstrip', *Alta Frequenza*, **45**, pp. 111–116

BRACKELMANN, W. (1967): 'Kapazitaten und induktivitäten gekoppelter streifenleitungen', *Arch. Elek Ubertragungstech*, **22**, pp. 313–321

CHANG, D. C., and KUESTER, E. F. (1979): 'An analytical theory for narrow-open microstrip', *Arch. Elek Ubertragungstech*, **33**, pp. 199–206

CORR, D. G., and DAVIES, J. B. (1972): 'Computer analysis of the fundamental and higher order modes in single and coupled microstrip', *IEEE Trans.*, **MTT-20**, pp. 669–678

DELOGNE, P. (1970): 'On Wu's theory of microstrip', *Electron. Lett.*, **6**, pp. 541–542

DENLINGER, E. J. (1971): 'A frequency dependent solution for microstrip transmission lines', *IEEE Trans.*, **MTT-19**, pp. 30–39

DEUTSCH, J., and JUNG, H. J. (1970): 'Measurement of the effective dielectric constant of microstrip lines in the frequency range 2 GHz to 12 GHz', *Nachrichtentech. Z.*, **23**, pp. 620–624

EASTER, B. (1975): 'The equivalent circuit of some microstrip discontinuities', *IEEE Trans*, **MTT-23**, pp. 655–660

FARRAR, A., and ADAMS, A. T. (1972): 'Matrix methods for microstrip three-dimensional problems', *IEEE Trans.*, **MTT-20**, pp. 497–504

FUJIKI, Y., HAYASHI, Y., and ZUZUKI, M. (1972): 'Analysis of strip transmission lines by interaction methods', *Electron. Commun. Japan*, **55**, pp. 74–80

GARG, R., and BAHJ, I. J. (1978): 'Microstrip discontinuities', *Int. J. Electron.*, **45**, pp. 81–87

GETSINGER, W. J. (1973): 'Microstrip dispersion model', *IEEE Trans.*, **MTT-21**, pp. 34–39

GETSINGER, W. J. (1979): 'Microstrip characteristic impedance', *IEEE Trans.*, **MTT-27**, p. 293

GUNSTON, M. A. R. (1972): 'Microwave transmission line impedance data' (Van Nostrand Reinhold Co., London)

GUPTA, C., EASTER, B., and GOPINATH, A. (1978): 'Some results on the end effects of microstriplines', *IEEE Trans.*, **MTT-26**, pp. 649–652

GUPTA, K. C., GARG, R., and BAHL, I. J. (1979): 'Microstrip lines and slotlines' (Artech)

HAMMERSTAD, E. O. (1975): 'Equations for microstrip circuit design', Proc. 5th European Microwave Conference, Hamburg, pp. 268–272

HAMMERSTAD, E. O., and BEKKADAL, F. (1975): 'Microstrip handbook', ELAB report STF 44 A74169, The University of Trondheim, The Norwegian Institute of Technology

HARLAN HOWE (1974): 'Stripline circuit design' (Artech)

HARTWIG, C. P., MASSÉ, D., and PUCEL, R. A. (1968): 'Frequency dependent behaviour of microstrip', *IEEE G-MTT* Int. Micr. Symp., pp. 110–116

HORNSBY, J. S., and GOPINATH, A. (1969a): 'Fourier analysis of a dielectric-loaded waveguide with a microstrip', *Electron. Lett.*, **5**, pp. 265–267

HORNBY, J. S., and GOPINATH, A. (1969b): 'Numerical analysis of a dielectric loaded waveguide with a microstrip line-Finite difference methods', *IEEE Trans.*, **MTT-17**, pp. 684–690

HORTON, R., EASTER, B., and GOPINATH, A. (1971): 'Variation of microstrip losses with thickness of strip', *Electron. Lett.*, **7**, pp. 490–491

HOSSEINI, M. N., and SHURMER, H. V. (1978): 'Computer-aided design of microwave integrated circuits', *Rad. & Electron. Eng.*, **48**, pp. 85–88

ITOH, T., and MITTRA, R. (1973): 'Spectral domain approach, for calculating dispersion characteristics of microstrip lines', *IEEE Trans.*, **MTT-21**, pp. 496–499

ITOH, T., and MITTRA, R. (1974): 'A technique for computing dispersion characteristics of shielded microstrip lines', *IEEE Trans.*, **MTT-22**, pp. 896–898

ITOH, T., MITTRA, R., and WARD, R. D. (1972): 'A method for computing edge capacitance of finite and semi-finite microstrip lines', *IEEE Trans.*, **MTT-20**, pp. 847–849

JANSEN, R. H. (1978): 'High speed computation of single and coupled microstrip parameters including dispersion, high-order modes, loss and finite strip thickness', *IEEE Trans.*, **MTT-26**, pp. 75–82

KADEN, H. (1974): 'Advances in microstrip theory', Siemens Forsch. u. Entwickl-Ber, Bd 3, Nr 2, pp. 115–124

KOMPA, G. (1978): 'Design of stepped microstrip components', *Rad & Electron. Eng.*, **48**, pp. 53–63

KOWALSKI, G., and PREGLA, R., (1972): 'Dispersion characteristics of single and coupled microstrips', *Arch. Elek. Ubertragungstech*, **26**, pp. 276–280

KNORR, J. B., and TUFEKCIOGLU, A. (1975): 'Spectral domain calculation of microstrip characteristic impedance', *IEEE Trans.*, **MTT-23**, pp. 725-728

KRAGE, M. K., and HADDAD, G. I. (1972): 'Frequency dependent characteristics of microstrip transmission lines', *IEEE Trans.*, **MTT-20**, pp. 678-688

KUESTER, E. F. and CHANG, D. C. (1979): 'An appraisal of methods for computation of the dispersion characteristics of open microstrip', *IEEE Trans.*, **MTT-27**, pp. 691–694

KUHN, E. (1973): 'A mode-matching method for solving field problems in waveguide and resonant circuits', *Arch. Elektron. & Uebertragungstech.*, **27**, (12), pp. 511–518

MARCUVITZ, N. (1951): 'Waveguide handbook' (McGraw-Hill, New York)

MEHRAN, R. (1976): 'Frequency dependent equivalent circuits for microstrip right-angle bends, T-junctions and crossings', *AEU*, **30**, pp. 80–82

MENZEL, W., and WOLFF, I. (1977): 'A method for calculating the frequency dependent properties of microstrip discontinuities', *IEEE Trans.*, **MTT-25**, pp. 107–112

MIRSHEKAR-SYAHKAL, D., and DAVIES, J. B. (1979): 'Accurate solution of microstrip and coplanar structures for dispersion and for dielectric and conductor losses', *IEEE Trans.*, **MTT-27**, pp. 694–699

MITTRA, R., and ITOH, T. (1971): 'A new technique for the analysis of the dispersion characteristics of microstrip lines', *IEEE Trans.*, **MTT-19**, pp. 47–56

MORGAN, S. P. (1949): 'Effect of surface roughness on eddy current losses at microwave frequencies', *J. Appl. Phys.*, **20**, pp. 352–358

OWENS, R. P. (1976): 'Accurate analytical determination of quasi-static microstrip line parameters', *Rad. & Electron. Eng.*, **46**, pp. 360–364

PREGLA, R., and KOWALSKI, G (1974): 'Simple formulas for the determination of the characteristic constants of microstrip', *Arch. Electron. & Uebertragungstech.*, **28**, pp. 339–340

PUCEL, R. A., MASSÉ, P., and HARTWIG, C. P. (1968): 'Losses in microstrip', *IEE Trans.*, **MTT-16**, pp. 342–350 (correction in **MTT-16**, (12) 1968)

SCHMITT, H. J., and SARGES, K. H. (1971): 'Wave propagation in microstrip', *Nachrichtentech Z.*, **24**. pp. 260–264

SCHNEIDER, M. V. (1969a): 'Microstrip lines for microwave integrated circuits', *Bell Sys. Tech. J.*, **48**, pp. 1421–1444

SCHNEIDER, M. V. (1969b): 'Dielectric loss in integrated microwave circuits', *Bell Sys. Tech. J.*, **48**, pp. 2325–2332

SILVESTER, P., and BENEDEK, P. (1973): 'Microstrip discontinuity capacitances for right-angle bends, T-junctions and crossings', *IEEE Trans.*, **MTT-21**, pp. 341–346 (correction in **MTT-23**, 1975, p. 456)

THOMSON, A. F., and GOPINATH, A. (1975): 'Calculations of microstrip discontinuity inductances', *IEEE Trans.*, **MTT-23**, pp. 648–655

TRAUT, G. R. (1979): 'Clad laminates of PTFE composites for microwave antennas', Proc Workshop on Printed Circuit Antenna Technology, New Mexico State University, Las Cruces, pp. 27–1 to 27–17

VAN DE CAPELLE, A. R., and LUYPAERT, P. J. (1973): 'Fundamental- and higher-order modes in open microstrip lines', *Electron. Lett.*, 9, pp. 345–346

VAN DE CAPELLE, A. R., and LUYPAERT, P. J. (1974): 'An investigation of the higher order modes in open microstrip lines' Proc. V. Colloq. Microwave Commun. Vol III Budapest: Akademiai Kiado, pp. ET–23––ET–31

VAN HEUVEN, J. H. C. (1974): 'Conduction and radiation losses in microstrip', *IEEE Trans.*, MTT-22, pp. 841–844

VAN DER PAUW, L. J. (1976): 'The radiation and propagation of electromagnetic power by a microstrip transmission line', *Philips Res. Rep.*, 31, pp. 35–70

WELCH, J. D., and PRATT, H. J. (1966): 'Losses in microstrip transmission systems for integrated microwave circuits', *NEREM Rec.*, 8, pp. 100–101

WHEELER, H. A. (1942): 'Formulas for the skin effect', *Proc. IRE*, 30, pp. 412–424

WHEELER, H. A. (1964): 'Transmission properties of parallel wide strips by a conformal mapping approximation', *IEEE Trans.*, MTT-12, pp. 280–289

WHEELER, H. A. (1965): 'Transmission-line properties of parallel strips separated by dielectric sheet', *IEEE Trans.*, MTT-13, pp. 172–185

WHEELER, H. A. (1977): 'Transmission-line properties of a strip on a dielectric sheet on a plane', *IEEE Trans.*, MTT-25, pp. 631–647

WIESBECK, W. (1972): 'Calculation and measurement of surface current distribution in unscreened strip lines', *Nachrichtentech Z.*, 25, pp. 1–6

WOLFF, I. KOMPA, G., and MEHRAN, R. (1972): 'Calculation method for microstrip discontinuities and T-junctions', *Electron. Lett.*, 8, pp. 177–179

WU, T. T. (1957): 'Theory of the microstrip', *J. Appl. Phys.*, 28, pp. 299–302

ZYSMAN, G. I., and VARON, D. (1969): 'Wave propagation in microstrip transmission lines' IEEE G-MTT Int. Mic. Symp. Digest, pp. 2–9

Radiation mechanism of an open-circuit microstrip termination – fundamental design implications

It was concluded in the previous Chapter that little had been established in the literature, either by measurement or analysis, about the precise radiation mechanism of discontinuities in microstrip. Radiation is, of course, a highly undesirable effect in microstrip circuits and the priority in this respect has clearly been to avoid radiation rather than investigate its origin and behaviour. Lewin (1960) was perhaps the first to consider the analysis of radiation from microstrip discontinuities, although microstrip antenna arrays comprised of step discontinuities had already been reported several years previous by Gutton and Baissinot (1955). Very recently there has been an upsurge of interest in the analysis of various forms of microstrip antennas and these will be described in later Chapters. In this present Chapter, attention will be confined to the behaviour of an open-circuit microstrip termination. This element is of particular interest because it is amenable to analysis and enables a clearer understanding of the physical aspects of other more complicated discontinuities to be obtained. In addition, it can itself be utilised as a radiating element and in conjunction with similar elements can form the basis of practical antenna arrays. There is little doubt that the recent practical antenna designs have benefited enormously from investigations into the fundamental radiation mechanism taking place in microstrip structures and we will therefore describe the measurements and mathematical techniques that have been the means to this design end. The few previous analyses are examined in the light of more recent theoretical and experimental work and fundamental properties concerning the performance and limitations of microstrip antennas extracted. This Chapter also serves as an introduction to the somewhat intractable mathematical problems surrounding the analysis of more complicated forms of discontinuities such as T junctions, step transitions and feed systems for microstrip antennas which are encountered in later Chapters.

3.1 Comparison of estimates of radiation loss

If the current flowing on the conducting surface of an antenna is known, then the radiation pattern and radiated power can be calculated. This will be referred to here as the 'current method' in contrast to the 'aperture method' whereby the radiation characteristics are calculated from the aperture fields; the aperture being some free-space surface (possibly coincident with parts of the boundary of the antenna structure) that encloses the antenna. There are many variations in these two basic approaches such as converting the aperture fields to fictitious equivalent magnetic and electric current sources or interpreting the dielectric polarisation throughout a dielectric material as a fictitious current source. All these formulations are rigorous, provided that the exact form of current or aperture field is known or if they can be solved for as an intermediate step in the analysis. Wire antennas have been exhaustively treated by the current method (King, Mack and Sandler, 1968) which leads to an integral equation in the unknown current distribution. Microwave horn antennas have been readily calculated by the aperture method (Silver, 1949) by assuming that the mouth of the horn is the dominant aperture region. The horn is, in fact, more typical of the type of antennas of practical interest for which a rigorous analysis is intractable and approximations are necessitated at the expense of accuracy. There is no way of improving the accuracy of the aperture method by successive iteration (Schelkunoff, 1951) and there appears to be no evidence that the current method is superior in this respect if the current distribution cannot be precisely formulated. Results obtained by making simplifying assumptions about current and aperture distributions need careful interpretation particularly so for the more complicated forms of antenna configurations where the sources of radiation are not well defined. An example of the latter is the surface wave antenna comprised of a short length of open transmission line such as a dielectric-rod (James, 1967), (James, 1970), (Bach Anderson, 1971). For this antenna the excitation source also contributes to the radiation pattern.

The above discussion gives immediate insight into the possible approaches to the approximate calculation of the open-circuit microstrip termination, Fig 3.1, and the salient problems arising. It is evident that the accuracy of the results is very dependent on the simplifying assumptions and the latter need to be carefully examined. The fields of a microstrip line are unbounded as in the dielectric rod case and although it is physically obvious that the radiation emanates predominantly from the terminal zone, a variety of aperture surfaces can be chosen. The structure is further complicated by the substrate-to-air interface and the need to consider the radiation loss from the source exciting the microstrip line. Since the approximate fields associated with the guided wave in the microstrip line are known it is possible to obtain first-order equations for the current or aperture distributions and Lewin (1960) and Sobol (1971) have, respectively, obtained radiation estimates in this way. The exact equivalence of two rigorous formulations has recently been illustrated by Chuang *et al.* (1980) but ways of improving commonly used approximate calculations are not considered.

In the current method, Lewin (1960) derives the electric and magnetic fields (**E**, **H**) from the current distribution **J**

$$\mathbf{E} = \nabla^2 \boldsymbol{\pi} + k_0^2 \boldsymbol{\pi} \tag{3.1}$$

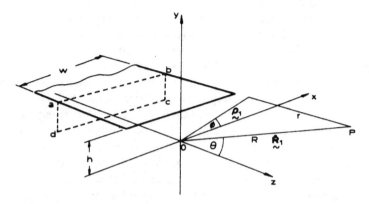

Fig. 3.1 *Systems of co-ordinates following notation of Silver (1949) associated with the open-circuit termination calculations eqns. 3.1–3.11*
In eqn. 3.2, $r = |R\hat{R}_1 - \rho_1|$

$$\pi = \frac{-j30}{k_0} \int \int \int_v \frac{e^{-jk_0 r}}{r} \, \mathbf{J} \, dv \tag{3.2}$$

where V is the volume occupied by the current $\mathbf{J}$ and r is defined in Fig 3.1. The current density $\mathbf{J}$ includes both the electric current in the strip and the fictitious electric current representing the dielectric polarisation of the substrate.

The simplifying assumptions are:

(i) The guided microstrip wave is taken as the quasi-TEM wave confined to the rectangular cross-section abcd of Fig 3.1.

(ii) The guided microstrip wave is completely reflected from the termination at $z = 0$, Fig 3.1.

(iii) The current in the strip flows only in the z direction, $J_x = 0, J_z \neq 0$.

(iv) Field and current relationships in the microstrip transmission line are calculated assuming the relative permittivity to be ϵ_0.

These assumptions restrict the applicability of the results to almost ideal open-circuit terminations where the radiation loss is negligible. Likewise the launching of substrate surface waves, which will be introduced in Section 3.3, is assumed to negligible.

Under these assumptions the strip current J_z is represented by

$$J_z = e^{-j\beta z} - e^{j\beta z} \tag{3.3}$$

From Maxwell's second equation the $(\mathbf{E}, \mathbf{H})$ fields are given by

$$\nabla \times \mathbf{H} = j\omega\epsilon_0\epsilon_e \, \mathbf{E} + \mathbf{J}$$

which may be rearranged as

$$\nabla \times \mathbf{H} = j\omega\epsilon_0 \mathbf{E} + [\mathbf{J} + \mathbf{J}_p] \tag{3.4}$$

$$J_p = j(\epsilon_e - 1) \omega\epsilon_0 E \tag{3.5}$$

where J_p is the fictitious equivalent electric current representing the dielectric polarisation. This current exists over the volume occupied by the quasi-TEM wave underneath the strip. The E-field in eqn. 3.5 is the E_y component of the quasi-TEM guided wave and the fictitious current is in the y-direction, thus

$$J_{py} = j(\epsilon_e - 1) \omega\epsilon_0 E_y \tag{3.6}$$

The total current J in eqn. 3.2 is the sum of J_z and J_{py} components of eqns. 3.3 and 3.6, respectively, and hence the E-field eqn. 3.1 is in principle obtained by relating E_y (eqn. 3.6) to J_z (eqn. 3.3) through the microstrip line impedance Z_m. The images of the currents in the ground plane are also embraced in eqn. 3.2. Unfortunately, both the strip current J_z and the fictitious volume current J_{py} extend over the entire semi-infinite length of the strip $-\infty \leqslant z \leqslant 0$ which is physically unacceptable and creates an improper integral requiring interpretation as a delta functional. Lewin gives a physical argument to deduce that the integral includes the source radiation at $z = -\infty$, or at $z = -L$ for a practically realisable length of line where $-L \leqslant z \leqslant 0$. On omitting the source term, a magnetic Hertzian dipole radiation pattern is obtained which corresponds to a radiated power P

$$P = 60(k_0 h)^2 F_1(\epsilon_e) \text{ watts}$$

$$F_1(\epsilon_e) = \left[\frac{\epsilon_e + 1}{\epsilon_e} - \frac{(\epsilon_e - 1)^2}{2\epsilon_e\sqrt{\epsilon_e}} \ln\left(\frac{\sqrt{\epsilon_e} + 1}{\sqrt{\epsilon_e} - 1} \right) \right] \tag{3.7}$$

Observing that this is peak power, the radiation conductance G_r is

$$G_r = 60 \left\{ \frac{\pi h}{Z_m \lambda} \right\}^2 F_1(\epsilon_e) \text{ Siemens} \tag{3.8}^\dagger$$

Sobol (1971) commences with the aperture approach and chooses as the aperture the $(xy0)$ plane, Fig 3.1. The result is obtained more readily than in the previous current method and the feed radiation is omitted as a matter of course. The underlying assumptions are similar to those invoked in the current method, items (i)–(iv) above, except that the permittivity ϵ_e is assumed throughout all space. The applicability is again restricted to near ideal open-circuit terminations with negligible loss of power to radiation and surface waves.

Under these assumptions, the aperture is taken at $z = 0$, and $-w/2 \leqslant x \leqslant w/2$ $-h \leqslant y \leqslant h$, the latter allowing for the image aperture in the groundplane. The magnetic field in the aperture is zero and the electric field $E(P)$ generated by the electric field E_T in the aperture is

$$E(P) \sim \int_{-w/2}^{w/2} \int_{-h}^{h} (\hat{z} \times E_T) \exp(jk_0 \sqrt{\epsilon_e}\, \rho_1 \cdot \hat{R}_1)\, dxdy \tag{3.9}$$

---

† The factor 120 in eqn. (1) of Wood, Hall and James (1978) requires division by 2 because Lewin (1960) defines power as peak and not mean. This has been implemented in Eqn (3.8).

where $\hat{R}_1$ is a unit vector in the direction of the observation point P in the far field and $\boldsymbol{\rho}_1$ is a vector from the origin 0 to a point in the aperture (Silver, 1949). Alternatively, the aperture electric field may be regarded as a layer of fictitious magnetic current and the fields at P derived from equations that are the magnetic current counterparts of eqns. 3.1 and 3.2.

This latter additional stage in the calculation was included by Sobol prior to obtaining the radiation conductance G_r

$$G_r = \frac{\sqrt{\epsilon_e}}{240\pi^2} F_2 \left(\sqrt{\epsilon_e} \frac{2\pi w}{\lambda_0} \right)$$

$$F_2(x) = xSi(x) - 2\sin^2\left(\frac{x}{2}\right) - 1 + \frac{\sin x}{x} \tag{3.10}$$

Both Lewin's and Sobol's analysis predict that the radiation pattern of the terminal zone is that of a Hertzian magnetic dipole but experimental confirmation of this is difficult because the source radiation has to be eliminated. James and Wilson (1977) have applied the aperture method to rectangular and cylindrical microstrip resonators and have also attempted to measure the radiation pattern of an isolated terminal zone with some success. Denlinger (1969) employed Lewin's equation (eqn. 3.7) to investigate the radiation power loss from microstrip resonators but an error in this work was pointed out by Easter and Roberts (1970) who reformulated the calculation to obtain an improved agreement with measurements. Belohoubek and Denlinger (1975) subsequently corrected Denlinger's previous calculation and also presented additional curves on the various Q factors for resonators. Roberts and Easter (1971) have also given measurements and estimates of the reduction in radiation loss that can be achieved from microstrip resonators having a U shape. Van Heuven (1974) gives a variety of measurements of the conduction and radiation losses in microstrip lines and resonators. He asserts that the theory of Sobol (1971) does not agree well with his measurements for silica and alumina substrates. Van Heuven (1974) finds that no firm conclusions can be drawn because the frequency dependence of the radiation is likely to be more involved than the theories suggest. Van Heuven however utilised the early results of Denlinger (1969) which as pointed out above have subsequently been corrected. Hammerstad and Bekkadal (1975) briefly review the above theories and propose that F_1 (ϵ_e) eqn (3.7) can be approximately taken as

$$F_1(\epsilon_e) \sim \frac{8}{3\epsilon_e} \tag{3.11}$$

since the treatment is already very approximate. They also consider that the treatment of Sobol (1971) includes parameters not embraced in that of Lewin (1960) and that it would be reasonable to take an average based on the two results of eqns. 3.7 and 3.10.

It is evident from the above discourse that considerable doubt has existed about the range of validity of Lewin's and Sobol's equations and Kompa (1976) asserts that the theories have been simplified to an inadmissable extent. Kompa (1976) points out that very wide open-circuit microstrip lines should tend to the behaviour of a semi-infinite dielectric loaded parallel plate waveguide which has been rigorously analysed by Angulo and Chang (1959). Measurements by Kompa (1976) confirm that this is so, but continue to leave open the question of the precise behaviour of the open-circuit termination of narrow lines.

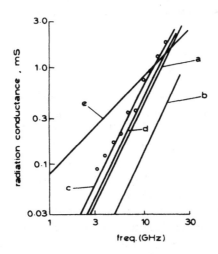

Fig. 3.2 *Comparison of G_r values obtained by various estimations*
   $w = 2 \cdot 82$ mm, $h = 1 \cdot 58$ mm, $\epsilon_r = 2 \cdot 32$
   a Lewin (1960)
   b Sobol (1971)
   c eqn. 3.10 using w_e instead of *w*
   d Surface field analysis eqn. 3.12 using w_e instead of *w*
   e Kompa (1976)
   Experimental o Wood *et al.* (1978)

A direct comparison of the above-mentioned theoretical estimates has been made by Wood, Hall, and James (1978) and many of the doubts resolved. The progress here is mainly due to the ability to measure the radiation conductance G_r directly and the measurement technique is described in the next Section. The comparative curves are given in Fig 3.2 and include an additional theoretical estimate, referred to as the 'surface field analysis' method, whereby the radiation is considered as originating from the electric field distribution at the substrate surface normal to the end of the line. Because of the presence of the ground plane, this component of field is only significant in the immediate vicinity of the line, $z > 0$ (Fig 3.1) and the radiation calculation may then be carried out in terms of an equivalent magnetic current. Allowing for the image in the ground plane, the expression for the radiated far field is

$$E = j \frac{\exp(-jk_0 R)}{2\lambda_0 R} \hat{R}_1 \times L$$

where **L** is the electric radiation vector given by

$$L = -2 \int_{-w/2}^{w/2} \hat{x} \, V \exp(jk_0 x \cos \phi \sin \theta) dx$$

and V is the voltage at the open circuit end of the line.

Evaluation of the power radiated then allows the end conductance to be calculated as

$$G_r = \frac{1}{120\pi^2} \, F_2 \left(\frac{2\pi w}{\lambda_0} \right) \tag{3.12}$$

The curves of Fig 3.2 show that the estimate of Kompa (1976) gives poor agreement for narrower widths as anticipated, but the frequency dependence is not predicted for any width. All the other methods exhibit a frequency dependence in accord with the experimental observation but both Sobol's method (eqn. 3.10) and the surface field analysis method (eqn. 3.12) require the strip width to be taken as an equivalent width (eqn. 2.16, Chapter 2) to embrace the field fringing effects. Lewin's estimate (eqn. 3.7) gives good agreement with the measurements. It is concluded that all these analytical estimates can give realistic data on the radiation conductance provided the assumed aperture or strip current distributions are adequate approximations to the local field behaviour on and around the strip conductor. This will be discussed again in Section 3.4.

3.2 Direct measurement of the end admittance

As has been stated above, the ability to directly measure the end admittance of an open-circuit microstrip termination has been an important contribution in this research and an outline of the main measurement technique used is now given. The technique involves the measurement of the voltage standing-wave ratio (VSWR) S of an open-circuit microstrip line and is thus simple in principle. The end admittance $G_r + jB$ is then given by

$$G_r = \frac{S}{Z_m \left(S^2 \cos^2 \frac{\Phi}{2} + \sin^2 \frac{\Phi}{2} \right)}$$

$$B = \frac{(S^2 - 1) \sin \frac{\Phi}{2} \cos \frac{\Phi}{2}}{Z_m \left(S^2 \cos^2 \frac{\Phi}{2} + \sin^2 \frac{\Phi}{2} \right)}$$

where $S > 1$ and Φ = the angular shift in the null position from that of an ideal open circuit.

In reality, the measurement procedure must be set up with precision and precautions taken to ensure that the line field is accurately monitored without perturbing it or picking up radiation from the terminal zone. The block diagram Fig 3.3 shows the equipment utilised in conjunction with the sliding carriage assembly and screening. The sliding carriage construction is shown in the photograph of Fig 3.4.

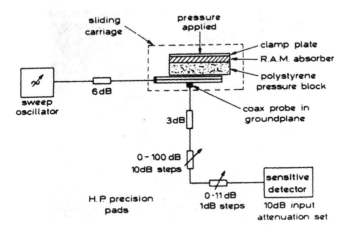

Fig. 3.3   *Block diagram of VSWR measurement jig*

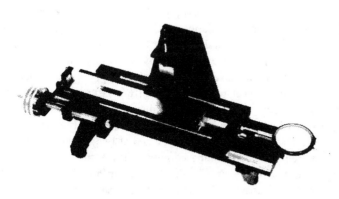

Fig. 3.4   *Photograph of VSWR jig showing travelling-probe assembly and precision rotary micrometer*

Preparation for the measurements commences by inserting a length of microstrip line on the ground plane of the sliding carriage. A 2 mm wide strip of copper has previously been etched from the copper backing of the microstrip substrate and this exposes the microstrip internal fields to the small probe which is flush mounted on a sliding plate in the carriage assembly. The photograph in Fig 3.5 shows a top view of the flush-mounted coaxial probe. The microstrip line is pressed onto the ground plane of the carriage with a block of polystyrene foam and with careful alignment

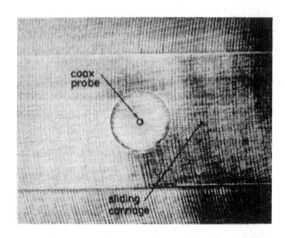

Fig. 3.5   *Enlarged photograph of sliding plate containing the flush-mounted coaxial probe*
The inner and outer diameters of the coaxial line are 0·38 mm and 0·73 mm, respec-
tively, and the diameter of the circular metal insert is 6 mm

the probe can be moved along the underside of the microstrip line with negligible disturbance because the slot in the microstrip ground plane is short-circuited by the base plate of the sliding carriage. Radiation absorbent material is placed in the far-field of the terminal zone on the metal surfaces of the carriage. The position of the minima in the line standing wave are indicated by the rotary micrometer and the ratio of maximum to minimum given on the probe output meter which utilises an existing facility within a spectrum analyser. Early measurements based on this principle (James and Wilson, 1977) yielded conductance values which had a large error spread for narrow lines. These errors have been considerably reduced by taking several amplitude readings about the minima and at two frequencies ±250 MHz about the nominal frequency. An online computer has been set up to automate this measurement and a sample shown in Fig 3.6. The influence on the VSWR of the surface waves, to be discussed in the next Section, was too small to be detected. However, it was considered useful to extend the substrate beyond the open circuit termination for a few centimetres to provide continuity to the microstrip ground plane.

So far, the method has been restricted to polyguide type substrates. Other ways of assessing G_r using resonator methods are mentioned in Section 3.4 and a curve fitting method based on the measurement of a linear combline array is mentioned in Chapter 5 section 5.4.3.

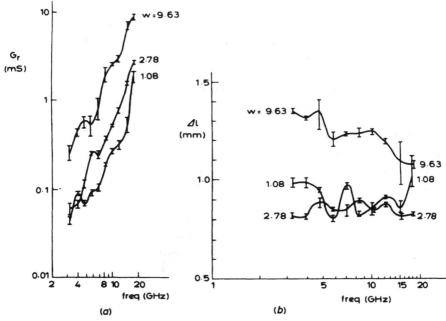

Fig. 3.6   *Measurements of end admittance using VSWR jig showing estimate of errors encountered; w is in mm, and $\epsilon_r = 2\cdot32$*
a G_r
b $\Delta l = (1/\beta) \tan^{-1} (\beta Z_m)$

3.3 Substrate surface waves and their significance

The use of microstrip as an antenna structure has demanded a deeper knowledge of
the behaviour of microstrip discontinuities and substrate surface waves are an issue
that has only relatively recently been considered (James and Ladbrooke 1973). Improved estimations of the power launched as a surface wave were subsequently
given by James and Wilson (1977). The concept of substrate surface waves is
immediately evident when one considers the fact that the dielectric substrate
together with its ground plane is capable of supporting dielectric slab-type modes
(Collin, 1960). Thus at any discontinuity in the conducting strip of a microstrip
line, and particularly at an open-circuit termination, the substrate surface waves will
be launched to some degree. The surface waves will cause isolated microstrip circuits
to become coupled but it turns out that the effect is generally second order for
circuit applications and in any case it is very difficult to measure. For microstrip
antenna applications, substrate surface waves impose a severe limit on the operation
at higher frequencies and result in uncontrollable coupling between radiating elements and scattering of radiation as the substrate waves impinge on discontinuities
in their path, i.e. the edge of the microstrip assembly.

From a theoretical standpoint the existence and behaviour of these substrate surface waves is well understood. Recent theoretical contributions which illustrate the interaction between the microstrip and substrate wave structures are Van der Pauw (1977), Ermert (1979) and Chang and Kuester (1979). For design purposes estimates of the degree of surface-wave launching are useful, and two methods are noted here. Kompa (1976) points out that the behaviour of an open-circuit termination in a wide microstrip line should tend to that of an infinitely wide line as rigorously analysed by Angulo and Chang (1959). On this basis, Kompa (1976) estimates that at 12·4 GHz an open-circuit microstrip line on a polyguide substrate ($\epsilon_r = 2·32$, $h = 1·58$ mm) will launch 4·5% of its incident power into substrate surface waves. If the substrate were alumina ($\epsilon_r = 9·7$ and $h = 0·635$ mm) the surface-wave power launched would only be 0·35%. The difficulty with this estimation is that it will not hold for narrow lines and no indication of how the accuracy is related to the line width is obtainable.

James and Wilson (1977) consider an open-circuit microstrip line mounted on a parallel-sided infinitely long substrate of width w_S (Fig 3.7). Modal representations of the substrate surface wave field ($\mathbf{E_s}$, $\mathbf{H_s}$) are taken and the microstrip incident field ($\mathbf{E}$, $\mathbf{H}$) is approximated by that of a quasi-TEM wave flowing directly under the strip. The fraction T of incident power P launched as substrate surface-wave power P_s is given by

$$T = \frac{1}{16P_sP} \left| \int_{s_a} (\mathbf{E_s} \times \mathbf{H} - \mathbf{E} \times \mathbf{H_s}).\, \hat{z}\, ds \right|^2 \tag{3.13}$$

where S_a is the rectangular interface aperture $-w/2 \leqslant x \leqslant w/2$, $0 \leqslant y \leqslant h$ at $z = 0$.
The result is

$$T = \frac{w}{w_sF} \left[\left\{ k_0 \left(\frac{\epsilon_r}{\kappa\beta\beta_s} \right)^{1/2} + \frac{1}{k_0} \left(\frac{\kappa\beta\beta_s}{\epsilon_r} \right)^{1/2} \right\} \left\{ \frac{\sin(k_2h)}{k_2h} \right\} \right]^2$$

$$F = \left\{ 1 + \frac{\sin(2k_2h)}{2k_2h} \left[(1 + \epsilon_r^2 \cot^2(k_2h)) \right] \right\} \tag{3.14}$$

where β_s is the phase constant of the surface-wave mode,

$$\beta_s^2 = \epsilon_r k_0^2 - k_2^2$$

and κ can be taken as approximately unity.

The result of eqn 3.14 for some common substrate configurations is given in Fig. 3.7 and indicate a lower degree of launching than the estimate of Kompa (1976). The limitation of eqn. 3.14 is the fact that it is only likely to give a reasonable estimation of T when w_s is not much larger than w, thus maintaining the rectangular duct nature of the substrate extension. When $w_s \gg w$, the surface waves will emanate from the termination as radial cylindrical waves and the slab mode upon which eqn. 3.14 is based will not be established until the cylindrical waves have been reflected from the distant edges of the substrate extension. In practice, the microstrip termination is surrounded by substrate and substrate waves will also be launched in the reverse direction.

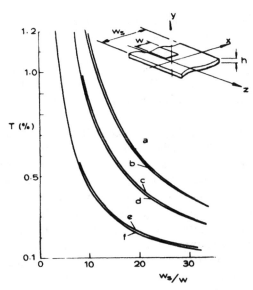

Fig. 3.7   *Computed values of T (eqn. 3.14) for various substrates and microstrip line im-pedances*

$$
\begin{array}{ll}
a\ Z_m = 100\,\Omega \\
b\ Z_m = \ \ 25\,\Omega
\end{array}\Bigg\}\quad \epsilon_r = 2.31,\ h = 1.58\ \text{mm at 10 GHz}
$$

$$
\begin{array}{ll}
c\ Z_m = 100\,\Omega \\
d\ Z_m = \ \ 25\,\Omega
\end{array}\Bigg\}\quad \epsilon_r = 2.31,\ h = 0.793\ \text{mm at 17 GHz}
$$

$$
\begin{array}{ll}
e\ Z_m = \ \ 50\,\Omega \\
f\ Z_m = \ \ 25\,\Omega
\end{array}\Bigg\}\quad \epsilon_r = 10,\ h = 0.635\ \text{mm at 17 GHz}
$$

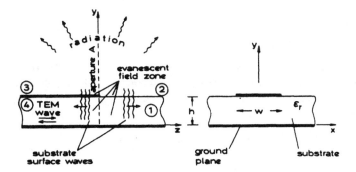

Fig. 3.8   *Sectional diagram of microstrip termination showing coordinates, parameters and field regions used with eqns. 3.15–3.26*

3.4 Improved analysis of microstrip open-circuit terminations

Three effects associated with a microstrip open-circuit termination have been identified having important design implication for both circuits and antennas. These effects are the well known end-effect Δl and the less familiar terminal radiation loss and substrate surface waves. The estimates of these effects so far are generally limited to low-frequency operation and each effect has been expressed in isolation from the other two. A more precise assessment of the behaviour of this microstrip termination over a wide frequency range is required and this has been recently addressed by James and Henderson (1979). A brief outline of the analysis using a variational method is now given and the design implications are considered in the next section.

The microstrip termination (Fig 3.8) is subdivided into four field regions for the purpose of the analysis. The boundaries of these regions are the parallel surfaces of the microstrip assembly and a vertical aperture A in the (xyo) plane. Incident and reflected guided waves travel in region 4, while radiation emanates into regions 2 and 3. Substrate surface waves are launched at the termination and travel radially outwards in both the positive and negative z directions. Near to the termination, the fields are evanescent creating a storage of energy and a subsequent susceptive loading B on the microstrip open-circuit line, where $\beta\Delta l = \tan^{-1}(BZ_m)$. The power lost to radiation and substrate surface waves manifests itself as end conductances G_r and G_s, respectively. The aim of this analysis is therefore to solve for the terminal admittance $Y = G_r + G_s + jB$ in the first instance.

In regions 1 and 2, the field distributions are characterised by Cartesian Hertzian electric and magnetic vector potentials $\pi_{E\frac{1}{2}}$ and $\pi_{H\frac{1}{2}}$, respectively, where

$$\pi_{E1} = a \sin(k_{y1}y) \cos(k_{x1}x) \exp(-j\beta_r z) \hat{z}$$

$$\pi_{E2} = \{c \sin(k_{y2}y) + d \cos(k_{y2}y)\} \cos(k_{x2}x) \exp(-j\beta_r z) \hat{z} \qquad (3.15)$$

where a, c and d are modal coefficients to be determined.

$$k_{x1}^2 + k_{y1}^2 = k_0^2 \epsilon_r \mu_r - \beta_r^2$$

$$k_{x2}^2 + k_{y2}^2 = k_0^2 - \beta_r^2 \qquad (3.16)$$

β_r is the spectral propagation constant and is real for radiation fields and imaginary for evanescent fields. Similar expressions exist for $\pi_{H\frac{1}{2}}$ in terms of additional unknown coefficients b, e and f. The components $E_{\frac{1}{2}}$ and $H_{\frac{1}{2}}$ of the field distributions are

$$E_i = \nabla(\nabla.\pi_{Ei}) + k_0^2 \epsilon_i \mu_i \pi_{Ei} - j\omega\mu_0 \mu_i \nabla \times \pi_{Hi}$$

$$H_i = j\omega\epsilon_0 \epsilon_i \nabla \times \pi_{Ei} + \nabla(\nabla.\pi_{Hi}) + k_0^2 \epsilon_i \mu_i \pi_{Hi}; \qquad i = 1, 2 \qquad (3.17)$$

On invoking the boundary conditions at the dielectric-to-air interface it is found that $k_{x1} = k_{x2} = k_x$ and c, d, e and f are expressible in terms of a and b. On introducing a second set of fields with coefficients $\bar{a}, \bar{b}, \bar{c}, \bar{d}, \bar{e}$ and $\bar{f}$ and invoking orthogonality between the two sets according to

$$\hat{z}.\int_{-\infty}^{\infty} \int_{0}^{\infty} E_i(k_x, k_{yi}) \times \bar{H}_i(k_x' k_{yi}') \, dy dx$$

$$= P(a, \bar{a}, b, \bar{b}) \, \delta\left(\frac{k_x}{k_0} - \frac{k_x'}{k_0}\right) \delta\left(\frac{k_{y2}}{k_0} - \frac{k_{y2}'}{k_0}\right); \qquad i = 1, 2 \qquad (3.18)$$

where δ is the delta function; it can be shown that all the coefficients can be determined as ratios of one of the coefficients. The radiation and evanescent fields of regions 1 and 2 are integrals of eqn. 3.17 over the appropriate range of the spectral propagation constant β_r which involves integrals over the spectral variables k_{x2} and k_{y2}. For the cases of interest $\epsilon_1 = \epsilon_r > 1$ and $\mu_1 = \mu_2 = \epsilon_2 = 1$. The substrate surface waves for regions 1 and 2 are characterised by Hertzian vectors $\boldsymbol{\pi}_{E\frac{1}{2}}^{s}$ and $\boldsymbol{\pi}_{H\frac{1}{2}}^{s}$ where

$$\boldsymbol{\pi}_{E1}^{s} = a_s \sin(k_{y1}y) \cos(k_x x) \exp(-j\beta_s z)\, \hat{z}$$

$$\boldsymbol{\pi}_{H1}^{s} = b_s \cos(k_{y1}y) \sin(k_x x) \exp(-j\beta_s z)\, \hat{z} \qquad (3.19)$$

a_s and b_s are unknown coefficients and $\boldsymbol{\pi}_{E2}^{s}$ and $\boldsymbol{\pi}_{H2}^{s}$ are given by a similar equation involving additional unknown coefficients c_s and e_s; k_{y1}, k_x and β_s satisfy a similar equation to eqn. 3.16 but this time k_{y1} and k_{y2} have discrete values which characterise the surface wave modes. k_x and β_s are again spectral values but it is found that

$$\beta_s = k_0 n_s \sin \alpha_s$$

$$k_x = k_0 n_s \cos \alpha_s$$

where $1 \leqslant n_s \leqslant (\epsilon_r)^{1/2}$, $k_{y1}^2 = k_0^2 (\epsilon_r - n_s^2)$, $k_{y2}^2 = k_0^2 (n_s^2 - 1)$. It can be shown that n_s is the discrete eigenvalue of a slab-type surface wave mode incident on the aperture A at an angle α_s. On applying the boundary conditions at $y = h$ the unknown coefficients can be determined as ratios of one coefficient and the modal solutions for the slab-type surface waves can be extracted. Only the first TM type mode having no cut-off is of interest for the substrate thickness and permittivities of interest. A typical mode chart for the substrate surface waves is shown in Fig. 3.9. It was found that the analysis became intractable when complete field representations were used for regions 3 and 4 and the following approximations were made.

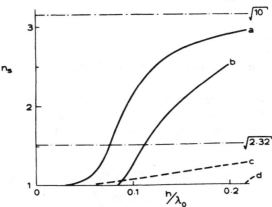

Fig. 3.9   *Mode diagram for substrate surface waves following the notation of Collin (1960)*

a TM$_0$ even, no cut-off   ⎫
b TE$_0$ odd, cut-off, $h/\lambda_0 = 0.084$ ⎬ $\epsilon_r = 10$
c TM$_0$ even, no-cut off   ⎫
d TE$_0$ odd, cut-off, $h/\lambda_0 = 0.216$ ⎬ $\epsilon_r = 2.32$

The microstrip fields were taken as the quasi-TEM wave:

$$E_y = \exp(-j\beta z) + \Gamma \exp(j\beta z)$$

$$H_x = -\left(\frac{\epsilon_e \epsilon_0}{\mu_0}\right)^{1/2} [\exp(-j\beta z) - \Gamma \exp(j\beta z)] \tag{3.20}$$

for $0 \leqslant y \leqslant h$ and $-w_e/2 \leqslant x \leqslant w_e/2$ in the region 4, where Γ is the reflection coefficient and w_e is the effective width of the line given by eqn. 2.16. The fields in region 3 are partially allowed for by the choice of the trial field E_A where

$$E_{xA} = 0, \quad E_{yA} = 1 \quad \text{for} \quad 0 \leqslant y \leqslant h, \quad -\frac{w_e}{2} \leqslant x \leqslant \frac{w_e}{2}$$

$$E_{xA} = E_{yA} = 0 \quad \text{elsewhere in the aperture } A \tag{3.21}$$

This choice of E_A effectively imposes a conducting plane in the aperture where $E_{yA} = 0$ and, provided $h \leqslant \lambda_0$, the effect is to double G_r. On equating the electric field components of eqns. 3.15, 3.19 and 3.20 to the trial field eqn. 3.21 and subsequently matching the corresponding magnetic fields, the variational solution for G_r is

$$G_r Z_m = \frac{1}{2} \int_0^{\pi/2} \int_0^{\pi/2} \frac{8[k_0^2 \epsilon_r^2 m k_{y1}^2 + \beta^2 k_x^2 g + 2\omega\epsilon_0 \epsilon_r n \beta k_x k_{y1}] f_e(A) d\gamma d\alpha}{\pi^2 \omega\epsilon_0 \sin(\alpha) k_0 h w_e (gm\mu_0/\epsilon_0 - n^2)(\epsilon_e \epsilon_0/\mu_0)^{1/2}}$$

$$\tag{3.22}$$

where

$$ga\bar{a} + m\frac{\mu_0}{\epsilon_0}b\bar{b} + \eta(a\bar{b} + \bar{a}b) = 0$$

and

$$f_e(A) = \left[\frac{\sin\left(k_x \dfrac{w_e}{2}\right)\sin(k_{y1}h)}{k_x k_{y1}}\right]^2$$

The reader is referred to James and Henderson (1979) for definitions of γ, α, m, n and g.

Integral expressions for both B and G_s are similarly obtained but are less useful because they are affected more severely by the approximations. After further consideration of the approximations the following forms reasonably model the behaviour of the termination

$$BZ_m = \frac{8}{\pi^2} \int_0^\infty \int_0^{\pi/2} \frac{f_e(A) \epsilon_r^{3/2} \mu_0 m k_{y1}^2}{\cos(j\alpha) h w\epsilon_0 \left(gm\dfrac{\mu_0}{\epsilon_0} + |\eta|^2\right)} d\gamma d\alpha \tag{3.23}$$

$$\frac{G_r}{G_s} = \frac{\cos^2(k_{y1}h)\epsilon_r k_{y1}^2}{(\cos^2(k_{y1}h)\epsilon_r k_{y2}^2 + k_{y1}^2 h k_{y2})} \tag{3.24}$$

Some computed results of G_r are given in Fig 3.10 (a) and (b), and it is seen that G_r rises to a peak value as the frequency increases. The peak is in the region where substrate surface waves are highly trapped and occurs approximately at $h/\lambda_0 = \frac{1}{4}/(\epsilon_r)^{1/2}$. When both h/λ_0 and $w_e/\lambda_0 \ll 1$, eqn. 3.22 simplifies to

$$G_r Z_m \sim \frac{4\pi}{3}\left(\frac{h}{\lambda_0^2} \; w_e\right) \frac{1}{(\epsilon_e)^{1/2}} \tag{3.25}$$

and this together with eqn. 3.8 is also plotted in Fig 3.10 (a) and (b) giving good agreement with eqn. 3.22 at lower frequencies.

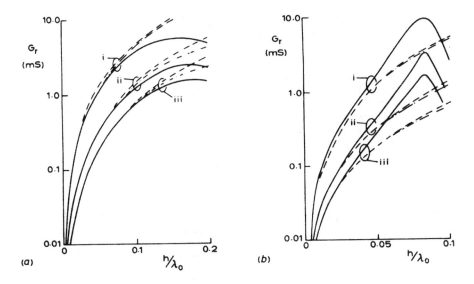

Fig. 3.10   *Computed G_r values*
a $\epsilon_r = 2\cdot32$
b $\epsilon_r = 10$
i w/h = 5
ii w/h = 1·79
iii w/h = 1
———— eqn. 3.22
· · · · · · eqn. 3.25
—·—·— eqn. 3.8

The effect of combining G_r and G_s is shown in Fig 3.11 (a) and (b), and the agreement with experimental data previously given by Wood, Hall and James (1978) is good. A variety of experimental and theoretical data are compared with Δl calcu-

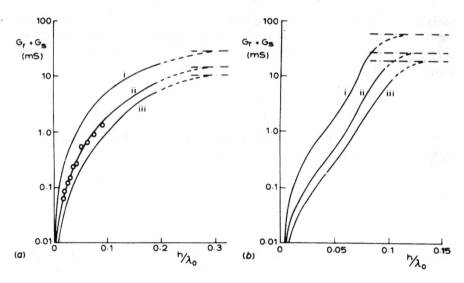

Fig. 3.11   *Computed* $(G_r + G_s)$ *values*

   a $\epsilon_r = 2 \cdot 32$, o o *measurements Wood et al. (1978)*
   b $\epsilon_r = 10$
   ——— *eqns. 3.22 and 3.24*
   · · · · · *extrapolation to asymptotic value of* $(1/Z_m)$
   i $w/h = 5$
   ii $w/h = 1 \cdot 79$
   iii $w/h = 1$

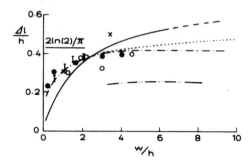

Fig. 3.12   *Comparison of* $\Delta l/h$ *values for alumina substrates*

   ——— *eqn. 3.23,* $h/\lambda_0 = 0 \cdot 02$, $\epsilon_r = 10$
   —··—··— *Itoh (1974),* $\epsilon_r = 9 \cdot 6$, $h/\lambda_0 \sim 0 \cdot 01$
   —··—·— *quasistatic, Silvester and Benedek (1972),* $\epsilon_r = 9 \cdot 6$
   · · · · · · *James and Tse (1972)* $\epsilon_r = 10$
   ll *Easter et al. (1978)* $\epsilon_r = 9 \cdot 8$, 8 GHz
   •••••• *James and Tse (1972),* $\epsilon_r = 10$ up to 12 GHz
   oooooo *Jain et al. (1972),* $\epsilon_r = 10$, 2 GHz
   xxxxxx *Troughton (1971)* $\epsilon_r = 9 \cdot 6$, 5 GHz

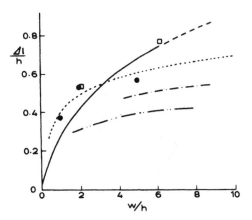

Fig. 3.13   *Comparison of Δl/h values for polyguide substrates*
    ———— eqn. 3.23 $h/\lambda_0 = 0\cdot02$, $\epsilon_r = 2\cdot32$
    —·—·—· quasistatic, $\epsilon_r = 2\cdot32$, empirical formula of Hammerstad and Bekkadal
    (1975)
    —··—··— Itoh (1974), $\epsilon_r = 2\cdot65$, $h/\lambda_0 \sim 0\cdot01$
    ······ quasistatic James and Tse (1972), $\epsilon_r = 2\cdot5$
    ●●●●●● James and Tse (1972), $\epsilon_r = 2\cdot53$ up to 12 GHz
    □□□□□□ $\epsilon_r = 2.3$, $h/\lambda_0 = 0.053$ by present authors

culated from B eqn. 3.23 in Figs. 3.12 and 3.13. There is much disagreement between the data due mainly to the difficulty of measurement. The actual behaviour of Δl as the frequency increases is shown in Fig. 3.14. A useful check on the accuracy of the present analysis is afforded by the transformation

$$G(\omega_1) - G(\omega_\infty) = -\frac{2}{\pi} \int_0^\infty \frac{\omega B(\omega) - \omega_1 B(\omega_1)}{\omega^2 - \omega_1^2} \, d\omega$$

$$B(\omega_1) = \frac{2\omega_1}{\pi} \int_0^\infty \frac{G(\omega) - G(\omega_1)}{\omega^2 - \omega_1^2} \, d\omega \qquad (3.26)$$

where $0 \leqslant \omega_1 \leqslant \infty$ and $G(\omega_\infty)$ denotes the value of $G = G_r + G_s$ at infinite frequency. This is also illustrated in Fig. 3.14, and shows that the analysis is reasonably consistent in the portrayal of the frequency behaviour of the end admittance. No way of measuring G_s has been found; some degree of comparison is possible however using data on surface-wave generation from patch resonators, Section 9.5.

3.5 Fundamental design implications based on first-order patch design

In Chapter 1 the bandwidth, efficiency and sidelobe levels of microstrip antennas were identified as performance factors that may be problematical for some applications. It was deduced that the impaired performance in these respects is the sacrifice necessary to achieve the very thin planar microstrip antenna assembly. The

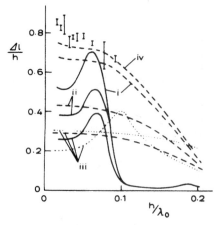

Fig. 3.14   *Variation of $\Delta l/h$ with frequency for finite width microstrip open-circuit termi-*
nations computed from eqn. 3.23

    ――――― $\epsilon_r = 10$, – – – $\epsilon_r = 2\cdot32$

i $w/h = 5$
ii $w/h = 1\cdot79$
iii $w/h = 1$
iv $w/h = 6\cdot1$
II experimental results for iv by present authors
. . . . . result of applying eqn. 3.26 to $(G_r + G_s)$
given in Figs. 3.11 (a) and (b)

above analysis for the microstrip open-circuit termination now enables the degree
of impairment to be usefully quantified for rectangular patch resonator antennas
and arrays of the latter.

A rectangular patch resonator consists of a microstrip line q half-wavelengths
long ($q = 1, 2, 3 \ldots$), and for antenna applications the case of most interest is $q =$
1. In practice some means of exciting the resonator either by attaching a feeder or
proximity coupling line at some appropriate impedance level on the resonator, is
required. Several patch resonators can be used for forming an array if some form of
feed network is provided but this aspect and the means of excitation is excluded in
this present assessment. The radiation pattern of a patch resonator can be considered
to be due to the combined radiation from each of the resonator ends. Even if
mutual coupling is ignored this provides a good first order radiation model and has
been derived by James and Wilson (1977). They show that the radiation patterns in
the principal orthogonal planes $\phi = \pi/2$ and $\phi = 0$ are, with reference to co-ordinates
of Fig 3.1 :

$$|E_\theta|^2 \sim \sin^2 \alpha\,(\theta)\,; \qquad \phi = \pi/2$$

$$|E_\phi|^2 \sim \left[\frac{\sin\left(\dfrac{k_0 w \sin\theta}{2}\right)}{\dfrac{k_0 w \sin\theta}{2}}\right]^2 \cos^2\theta \, \sin^2 \alpha(\theta)\,; \qquad \phi = 0 \qquad\qquad (3.27)$$

where

$$\alpha(\theta) = \frac{2\pi l}{\lambda_g} \left(\frac{\cos\theta}{(\epsilon_e)^{1/2}} - 1 \right) \qquad 0 \leqslant \theta \leqslant \pi/2$$

and w and $2l$ are the width and length of the microstrip patch resonator, respectively. The sin $\alpha(\theta)$ factor accounts for the interference pattern between the two end sources and the $(\sin x)/x$ function only contributes for very wide patches. Apart from these factors the pattern is essentially dipole-like and is dominated by the $\cos^2\theta$ term. More refined methods of analysis will be introduced in later Chapters, but for the purposes of this assessment it is sufficient to assume that the power radiated by the patch resonator creates a predominantly dipole-like pattern given by eqn. 3.27. A more precise analysis of the behaviour of the radiation and surface wave fields at the substrate surface and an assessment of coupling between resonators has been given by James and Henderson (1979). The question addressed here is, however, how much of the available power supplied to the resonator is lost both as heat and to the generation of substrate surface waves. To assess the former we note that the Q factor of the resonator Q_T can be expressed as

$$\frac{1}{Q_T} = \frac{1}{Q_c} + \frac{1}{Q_d} + \frac{1}{Q_R} \tag{3.28}$$

where Q_c, Q_d and Q_R are the Q factors corresponding to the conductor, dielectric and radiation losses, respectively. The conductor and dielectric losses can be calculated from the formulas of Chapter 2, Section 2 and for the substrates of interest the conductor loss is by far the greater.

$$Q_R = \frac{q\pi|\Gamma|}{1-|\Gamma|^2} \tag{3.29}$$

where Γ is defined in eqn. 3.20. Neglecting the dielectric loss the efficiency η and antenna loss factor L_A are given by

$$\eta = \frac{Q_c}{Q_c + Q_R} \times 100\% ; \qquad L_A = 10 \log_{10}\left(1 + \frac{Q_R}{Q_c}\right) dB \tag{3.30}$$

Some examples of Q_T for polyguide and alumina resonators are given in Fig. 3.15 and the individual Q_c and Q_R factors are plotted in Fig. 3.16 (a), (b) and (c). The latter figures also include curves of S_L:

$$S_L = \frac{G_s}{G_r} = \frac{P_{SUR}}{P_{RAD}} \tag{3.31}$$

where P_{SUR} and P_{RAD} are the powers lost to surface waves and radiation, respectively, and G_s/G_r is defined in eqn. 3.24. S_L gives a coarse worst case indication of the sidelobe level limit because the power lost to surface waves could conceivably be reradiated at the substrate extremities and other discontinuities to produce unwanted radiation. This assumes that the reradiated surface wave power will add in a phase coherent manner which is unlikely to be so. Nevertheless, S_L is a useful indication which together with the other data in Fig. 3.16 quantifies the constraints on bandwidth, efficiency and sidelobe levels. Even more important it enables a comparison of substrates to be made as follows.

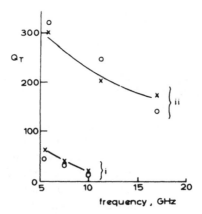

Fig. 3.15 *Q-factors of rectangular microstrip resonators*
oo experimental, James and Henderson (1979)
xx computed from eqns. 2.9, 2.12, 3.28 and 3.29 with $\sigma = 56 \times 10^4$ (S/cm)
i $\epsilon_r = 2.32$, $h = 1.58$ mm, $w = 4.65$ mm, $q = 1$
ii $\epsilon_r = 10$, $h = 0.625$ mm, $w = 0.6$ mm, resonator length is 9.37 mm, thus $q = 1, 2$
and 3.

In each figure of Fig. 3.16, S_L increases with frequency whereas L_A and η worsen with decreasing frequency. Since operation at extreme frequencies is therefore to be avoided the concept of a 'window' of permissible operating frequencies is suggested. For instance, the operating window shown in Fig. 3.16 (a) is defined by $\eta = 50\%$ and $S_L = -10$ dB for $w/h = 5$. In this case an antenna efficiency of better than 50% and sidelobe levels in an array of at least -10 dB are obtainable within the range 1·5–13 GHz. As stated before this does not include feeder losses or phase incoherence effects in the reradiated surface waves. In fact, experience indicates that S_L is very pessimistic and sidelobe levels probably of at least -20 dB would be obtainable in an array of patch elements. However, it is instructive to take S_L as it stands to compare the three substrates in Fig. 3.16. It is seen that halving the thickness of the substrate translates the window to about 3–35 GHz (Fig. 3.16 (b)) and for alumina (Fig. 3.16 (c)) the corresponding window is about 4–28 GHz. There would appear to be no advantage in using the alumina substrate unless a lower sidelobe level is specified. For instance, neither of the two polyguide substrates have S_L much lower than -20 dB, yet Fig 3.16 (c) shows that for alumina S_L values approaching -35 dB are obtainable below 10 GHz. It follows that thinner substrates composed of the same material cause the operating window to shift upwards in frequency whereas the use of material with higher permittivity leads to improved S_L values over a narrower operating range of frequencies. Finally, the bandwidth of the antenna over the operating window can be assessed from the Q values portrayed in Fig. 3.16. It is seen from Fig. 3.16 *a* and *b* that at a given frequency of operation the effect of reducing the substrate thickness is to reduce the resulting bandwidth. From this standpoint the thickest substrate consistent with the desired S_L value should be chosen. For the same substrate thickness the use of a higher permittivity

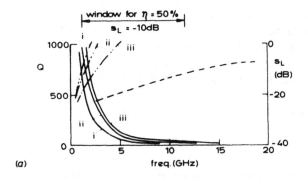

(a)

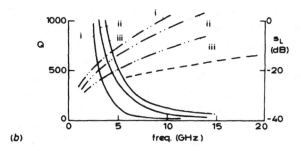

(b)

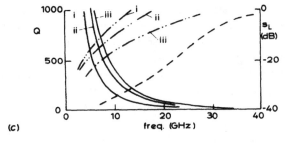

(c)

Fig. 3.16   *Computed microstrip antenna design curves*
  a $\epsilon_r = 2\cdot32, h = 1\cdot58$ mm
  b $\epsilon_r = 2\cdot32, h = 0\cdot79$ mm
  c $\epsilon_r = 10, h = 0\cdot64$ mm
  — — — S_L
  ——— Q_R, eqn. 3.29
  —··—··— $Q_c; \sigma = 56 \times 10^4$ (S/cm), eqn. 2.12
  i $w/h = 5$
  ii $w/h = 1\cdot79$
  iii $w/h = 1$

material again decreases the bandwidth and efficiency in exchange for lower S_L values. The deduction in Chapter 1 that bandwidth and efficiency must be traded for a thinner planar structure is well borne out by this data but the sidelobe level, pessimistically portrayed by S_L is a new parameter peculiar to microstrip antennas, making the assessment much more involved. It therefore seems likely that the upper limit to all types of microstrip antennas is where surface wave action prevents proper control of the aperture distribution and S_L gives a useful comparative measure of this effect.

3.6 Considerations for other antenna forms and discontinuities

The importance of the analysis of the microstrip open-circuit termination is that it enables actual design performance limitations to be assessed for rectangular resonator patch antennas. The analysis reveals the various aspects of the physical behaviour which are in any case difficult to isolate and measure in practice. The analysis suggests that if resonator action could be avoided then improved efficiency and bandwidth would result although the surface wave limitations at high frequency would remain. The need to leak out the radiation from a series of discontinuities is one possible solution and this is discussed in later chapters dealing with new types of travelling-wave antennas. The need to analyse various other microstrip discontinuities is therefore very apparent if precise control of the antenna design is to be achieved but in general the mathematical development becomes even more intractable. Circular patch resonators allow compatible cylindrical functions to be incorporated in the analysis and the unknown current in the conductor to be solved by moment methods using basis functions. This and other recent mathematical techniques will be dealt with in later Chapters. More general discontinuities such as corner bends, T junctions, steps and tapers remain essentially unsolved from the standpoint of obtaining precise design data about their radiation characteristics. Apart from Lewin (1978 a and b) which is a low-frequency estimation of various discontinuities neglecting surface wave effects, the remaining contributions are quasistatic analyses which neglect radiation as well; the latter have been discussed in Chapter 2. The difficult problem of calculating the radiation loss from the region where a microstrip antenna is attached to a cable is dealt with in Chapter 6.

3.7 Summary comments

Existing and more recent estimations and analyses of the microstrip open circuit terminations are outlined and compared. Also a direct means of measuring the terminal admittance based on a probe measurement of the standing waves on the line is discussed. The significance of substrate surface wave generation is a major conclusion in that it presents a high-frequency limit on the operation of microstrip antennas. As this limit is approached progressively more power is launched as surface waves which can subsequently reradiate at discontinuities and disrupt the operation of a microstrip termination as an antenna element; in particular, the side-lobe level of an array of such elements will be difficult to control at low levels. Unlike the radiation conductance and susceptance no suitable method of measurement for the surface waves has yet been evolved and the deductions here are based on the parameter S_L derived from the analysis. Consideration of conductor losses enables the Q factor of a rectangular resonator to be calculated thus giving information about efficiency and bandwidth of rectangular resonator type antennas which together with S_L quantifies the extent to which sidelobe level, bandwidth and efficiency have to be traded for the very thin planar structure. The concept of a window of operating frequencies for arrays of patch resonators on a given substrate is conceived. Although the microstrip open circuit is perhaps the simplest form of microstrip discontinuity its analysis is seen to lead to a major advancement in the understanding of the fundamental behaviour and limitations of microstrip antennas. As a result it is evident that if resonator action can be avoided by feeding the open-circuit termination in some other way then the efficiency and bandwidth limitations should be eased. This deduction is followed up in later Chapters and new types of broadband antennas result. Unfortunately these antennas have more complicated forms of discontinuities, i.e. bends and corners and the difficulty of analysing these has been emphasised in this present Chapter.

3.8 References

ANGULO, C. M., and CHANG, W. S. C. (1959): 'The launching of surface waves by a parallel plate waveguide', *IRE Trans.*, AP-7, pp. 359–368

BACH ANDERSEN, J. (1971): 'Metallic and dielectric antennas' (Polyteknisk Forlag)

BELOHOUBEK, E., and DENLINGER, E, (1975): 'Loss considerations for microstrip resonators', *IEEE Trans.*, MTT-23, pp. 522–526

CHANG, D. C., and KUESTER, E. F. (1979): 'An analytic theory for narrow open microstrip', *Arch. Elek. Ubertragungstech*, 33, pp. 199–206

CHUANG, S. L., TSANG, L., KONG, J. A., and CHEW, W. C. (1980): 'The equivalence of the electric and magnetic surface current approaches in microstrip antenna studies', *IEEE Trans*, AP-28, pp. 569–571

COLLIN, R. E. (1960): 'Field theory of guided waves', (McGraw-Hill) pp. 470–474

DENLINGER, E. J. (1969): 'Radiation from microstrip resonators', *IEEE Trans.* MTT-17, pp. 235–236

EASTER, B., and ROBERTS, R. J. (1970): 'Radiation, from half wavelength open-circuit microstrip resonators', *Electron. Lett.*, 6, pp. 573–574

EASTER, B., GOPINATH, A., and STEPHENSON, I. M. (1978): 'Theoretical and experimental methods for evaluating discontinuities in microstrips', *Rad. & Electron. Eng.*, 48, pp. 73–84

ERMERT, H. (1979): 'Guiding and radiation characteristics of planar waveguides', *IEE J. Microwave Optics & Acoustics*, 3, pp. 59–62

GUTTON, H., and BAISSINOT, G. (1955): 'Flat aerial for ultra high frequencies' French Patent No. 703113

HAMMERSTAD, E. O., and BEKKADAL, F (1975): 'Microstrip handbook' ELAB report STF44 A74169, University of Trondheim, pp. 77–82

ITOH, T. (1974): 'Analysis of microstrip resonators', *IEEE Trans.*, MTT-22, pp. 946–952

JAIN, O. P., MAKIOS, V., and CHUDOBIAK, W. J. (1972): 'Open-end and edge effects in microstrip transmission lines', *IEEE Trans.* MTT-20, pp. 626–628

JAMES, D. S., and TSE, S. H. (1972): 'Microstrip end effects', *Electron. Lett.*, 8, pp. 46–47

JAMES, J. R. (1967): 'Theoretical investigation of cylindrical dielectric-rod antennas', *Proc. IEE*, 114, pp. 309–319

JAMES, J. R. (1970): 'Applicability of Kirchoff integral to antenna near-fields', *Electron. Lett.* 6, pp. 547–549

JAMES, J. R., and HENDERSON, A. (1979): 'High frequency behaviour of microstrip open-circuit terminations', *IEE J. Microwave Optics & Acoustics*, 3, pp. 205–211

JAMES, J. R., and LADBROOKE, P. H. (1973): 'Surface wave phenomena associated with open-circuited stripline terminations', *Electron. Lett.*, 9, pp. 570–571

JAMES, J. R., and WILSON, G. J. (1977): 'Microstrip antennas and arrays Pt 1 – Fundamental action and limitations', *IEE J. Microwave, Optics & Acoustics*. 1, pp. 165–174

KING, R. W. P., MACK, R. B., and SANDLER, S. S. (1968): 'Arrays of cylindrical dipoles' (Cambridge University Press)

KOMPA, G. (1976): 'Approximate calculation of radiation from open-ended wide microstrip lines', *Electron. Lett.*, 12, pp. 222–224

LEWIN, L. (1960): 'Radiation from discontinuities in stripline', *Proc. IEE*, 107C, pp. 163–170

LEWIN, L. (1978a): 'Spurious radiation from microstrip', *Proc. IEE*, 125, pp. 633–642

LEWIN, L. (1978b): 'Spurious radiation from a microstrip Y junction', *IEEE Trans.*, MTT-26, pp. 895–896

ROBERTS, R. J., and EASTER, B. (1971): 'Microstrip resonators having reduced radiation loss', *Electron. Lett.*, 7, pp. 191–192

SCHELKUNOFF, S. A. (1951): 'Kirchhoff's formula its vector analogue and other field equivalence theorems', *Comm. Pure. Appl. Maths*, 4, pp. 43–59

SILVER, S. (1949): 'Microwave antenna theory and design' (McGraw-Hill) Chapter 10

SILVESTER, P., and BENEDEK, P. (1972): 'Equivalent capacitances of microstrip open circuits', *IEEE Trans.*, MTT-20, pp. 511–516

SOBOL, H. (1971): 'Radiation conductance of open-circuit microstrip', *IEEE Trans.*, MTT-19, pp. 885–887

TROUGHTON, P. (1971): 'Design of complex microstrip circuits by measurements and computer modelling', *Proc IEE*, 118, pp. 469–474

VAN HEUVEN, J. H. C. (1974): 'Conduction and radiation losses in microstrip', *IEEE Trans.*, MTT-22, pp. 841–844

VAN DER PAUW, L. J. (1977): 'The radiation of electromagnetic power by microstrip configurations', *IEEE Trans.*, MTT-25, pp. 719–725

WOOD, C., HALL, P. S., and JAMES, J. R. (1978): 'Radiation conductance of open-circuit low dielectric constant microstrip', *Electron. Lett.*, 14, pp. 121–123

Chapter 4

Basic methods of calculation and design of patch antennas

The simplest microstrip antenna configuration is the resonator or patch and this was briefly introduced in Section 1.4 and again in Section. 3.5 where some fundamental design implications were deduced. These previous Chapters are intended to provide foundation material and as such the microstrip antennas briefly mentioned so far have been treated with some generalisation. In this Chapter we now focus on the basic patch antenna; early developments, physical action, elementary theories, basic geometries of interest, realistic performances obtainable, refinements to theory and latest developments are addressed.

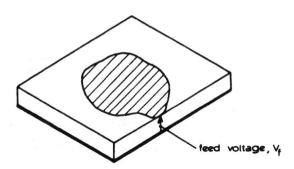

Fig. 4.1 *Microstrip patch antenna driven against ground plane*

   The patch antenna was among the earliest microstrip antennas developed, the structure and operation being related to that of resonators used in circuit design (Gupta *et al.*, 1979). The antenna consists of an isolated area of conductor on the upper surface of the microstrip substrate, with dimensions comparable to a half-wavelength, and is driven by a feed voltage V_f applied between the conductor and the ground plane of the microstrip, Fig. 4.1. Antennas of this configuration were described by Byron (1972), for a phased array application at 4·8 GHz and included disc and strip shaped resonators (Fig. 4.2 (a) and (b)). Although these antennas were constructed using a dielectric sheet to separate the radiators and the ground plane, printed circuit manufacture techniques were not used and the substrate thicknesses ($> 6\cdot4$ mm) were greater than those currently used with printed circuit microstrip antennas. The theoretical description of the physical action was given in terms of dipole radiation, which is not suitable to fully explain the radiation charac-teristics of this type of antenna.

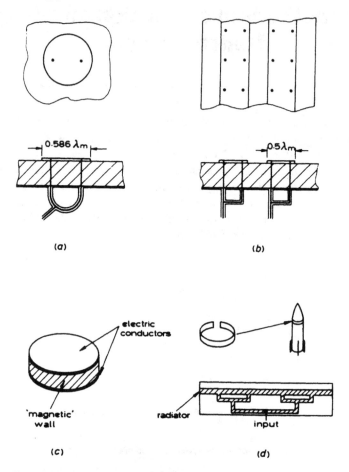

Fig. 4.2   *Microstrip patch antenna configurations*
    a Circular disc with symmetric coaxial feeds
    b Long strip radiator with multiple symmetric coaxial feeds
    c Magnetic wall cavity resonator model of patch antennas
    d Conformal long strip radiator with multiple microstrip feeds

Munson (1974) and Howell (1975), model the patch antenna as a cavity with high-impedance walls (Fig. 4.2c), with radiation occurring from the slot formed by the periphery of the radiator and the ground plane. This cavity model has formed the basis for the analysis of the microstrip patch antenna, and some of the refinements needed to improve the accuracy of this simple model for particular shapes of radiator are given in the following Sections of this Chapter.

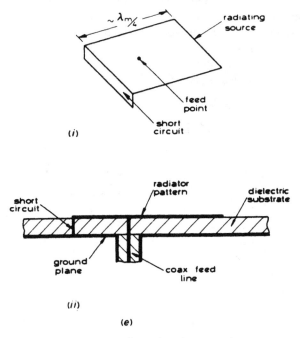

Fig. 4.2 *Microstrip patch antenna configurations (continued)*
    e Use of short circuits at patch edges

Further developments by Munson included the idea of wrapping the antenna around the body of a vehicle or missile as sketched in Fig. 4.2(d). Other ideas that followed concerned the use of short circuits within or along the patch boundary (Fig. 4.2(e)). The simplest way of creating a short circuit is to use a straight short-circuit plane with a rectangular patch geometry as described by Sanford and Klein (1978), but the concept can be applied with success to other more complicated forms. Garvin *et al.* (1977) describe the use of semicircular short-circuited patches in combination with a rectangular short-circuited patch to generate sum and difference radiation patterns from an antenna system mounted on the base of a missile or spacecraft. More recent work on short circuit patches is discussed in Section 4.6.

The excitation of patch antennas present both practical and analytical difficulties. A widely used configuration is a microstrip transmission line etched upon the same circuit board, connected at the edge of the patch (Fig. 4.3(a)) which has the advantage that circuit elements may be integrated upon the same printed circuit board (Munson, 1974), provided the extra space needed is available. A problem with this method, however, is that the patch has a high-radiation impedance located at the edge, as shown later in this Chapter; obtaining a good match to a useable size of transmission line feed may be difficult and requires an extensive matching network. This can be avoided if the feed voltage is applied at a point within the resonant field structure of the patch which is at lower voltage than that at the edge. An impedance transformation effect then occurs and a good match to any desired impedance may generally be obtained. Such a feed may be obtained (Howell, 1975) by running a coaxial line to a suitable point on the groundplane and connecting the inner through the substrate to the radiator whilst the outer is connected to the ground plane (Fig. 4.3(b)). This method, however, detracts from the simple printed-circuit board construction if more than one element is used in the antenna.

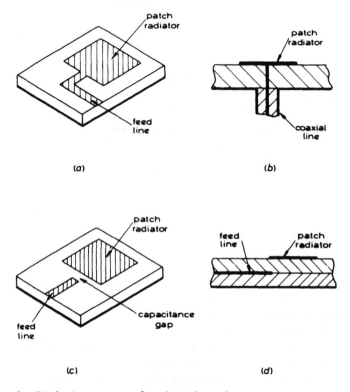

Fig. 4.3   *Possible feed arrangements for microstrip patch antennas*
a Surface microstrip transmission-line fed with direct connection
b Coaxial line feed through substrate
c Surface microstrip transmission line feed with matched capacitance coupling
d Double-layer microstrip transmission line feed

An impedance transformation may also be achieved by using capacitive coupling between the patch and line Fig. 4.3c, where the overall resonant frequency is determined by the inductive component of the patch edge impedance, matching that of the coupling capacitance. The frequency detuning thus produced leads to a reduced resonant impedance and a good match, but the coupling capacitance required is generally so small as to make design difficult and production totally unreliable. Variations on these ideas may be made by using a multilayer dielectric with the feed lines closer to the ground plane than the radiator, Fig. 4.3*d*. The feed may be directly coupled by a pin through the upper substrate (Garvin *et al.*, 1977) or by fringe-fields (Oltmann, 1978). This arrangement has the advantage of reducing unwanted feed radiation which may be a problem with single-layer substrates as discussed later in Chapter 6. It also allows the possibility of saving space by placing parts of the feed network underneath the patch radiators if coupling problems can be overcome. The increased constructional complexity of the antenna is obviously a problem.

Following these introductory details, Section 4.1 considers the fundamental aspects of patch analysis. Sections 4.2 and 4.3 describe the calculation of the radiation properties of rectangular and circular patches using the modal approach. Analysis of the input impedance is discussed in Section 4.4. Some refinements to the theoretical models used so far are introduced in Section 4.5, and finally the performance and analysis of short-circuit patches is described in Section 4.6.

4.1 Patch analysis

4.1.1 Field structure of patch
The field structure appropriate to a patch resonator is dependent on the relationship of the electric and magnetic fields at the edge of the patch. It is clear that the close spacing of the patch conductor to the ground plane, which is $< \lambda_0/10$ for typical microstrip substrates, will tend to concentrate the fields underneath the patch. The fields will leak out into the air through the substrate surrounding the patch, leading to a complex boundary condition problem. This problem may be approximated by applying an open-circuit boundary condition at the edges of a microstrip patch, a choice which may be justified by considering the behaviour of a TEM parallel-plate wave approaching the straight edge of a semi-infinite plate above a ground plane, Fig. 4.4. This problem is related to those having an exact Wiener-Hopf solution (Noble, 1958; Marcuvitz, 1951), the reflected wave within the plate system being given by

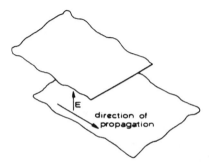

direction of
propagation

Fig. 4.4 *TEM wave incident on edge of semi-infinite parallel-plate system*

$$\Gamma = \exp\left(\frac{2\pi h}{\lambda_0}\right) \exp(-j\Phi) \tag{4.1}$$

where

$$\Phi = \frac{4h}{\lambda_0} \ln\left(\frac{e\lambda_0}{\gamma h}\right) - \sum_{n=1}^{\infty}\left\{\sin^{-1}\left(\frac{2h}{n\lambda_0}\right) - \frac{2h}{n\lambda_0}\right\}; \quad \begin{array}{l} e = 2\cdot718 \\ \gamma = 1\cdot781 \end{array}$$

The phase of the reflection coefficient is small for $h \sim 0\cdot01\,\lambda_0$, and the equivalent circuit of the system may then be represented by a shunt conductance located slightly outside the physical edge of the plate at a distance Δl, Fig. 4.5. The normalised magnitude of the conductance is $\sim 0\cdot03$ for $h \sim 0\cdot01\,\lambda_0$, showing that the edge behaves essentially as an open circuit located just outside its physical position.

Fig. 4.5   *Equivalent circuit parameters of parallel-plate edge*

Similar results were obtained by Angulo and Chang (1959) for the case where a dielectric slab is included in the plate system, which is then more representative of the microstrip patch problem, but the solution is more complicated than eqn. 4.1 and no details will be given here. However, this analysis has also been used by Kompa to characterise the reflection coefficient of very wide microstrip transmission lines (Section 3.1), and does appear to give agreement in the wide line limit with other theoretical and experimental results.

The high-impedance condition at the patch periphery means that the electric field parallel to the edge has a maximum whilst the magnetic field has a minimum. The first step in the analysis is to make the approximation that the magnetic field parallel to the edge is zero, and this is modelled by placing a wall of 'magnetic' conductor at the effective edge position, Fig. 4.6. The boundary conditions for the patch are expressed mathematically as

$$\mathbf{H}\cdot\mathbf{z} = 0, \mathbf{E}\times\mathbf{z} = 0; \qquad z = 0 \text{ and } h \tag{4.2a}$$

$$\mathbf{E}\cdot\mathbf{n} = 0, \mathbf{H}\times\mathbf{n} = 0; \qquad \text{on magnetic wall} \tag{4.2b}$$

The first-order approximation to the microstrip antenna is therefore an enclosed cavity, which will be able to support an infinite set of resonant modes of different frequencies. Since the height of the cavity is very small compared to a wavelength for operation as a microstrip antenna, the restriction may be imposed that only modes with $(\partial/\partial z)\,(\mathbf{E}, \mathbf{H}) = 0$ need to be considered. Combining this with the boundary condition (4.2a)

$$E_x = E_y = H_z = 0 \tag{4.3}$$

throughout the whole volume, independent of the patch shape.

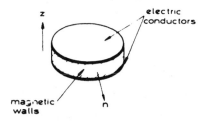

Fig. 4.6   *Magnetic wall model of patch antenna*

Under these conditions, the field structure within the cavity at resonance is the dual of the field structure at the cutoff-frequency of a TE mode in a metal wave-guide whose conducting boundary has the same shape as the effective boundary of the patch. Thus it is not normally necessary to analyse the field structures of this simplified model of the microstrip directly since the solution for most cases will already be available as the dual waveguide solution. One restriction which must be applied if a simple analysis is to be possible is that the patch shape should be such that the mode fields are derivable as simple expression. Lo *et al.* (1979) give the field expressions for a number of patch shapes; however there is little advantage to be gained by using configurations other than rectangular or circular forms and the examples dealt with here will be restricted to these two cases.

In a practical device, the feed connected to the patch antenna will excite all resonant modes of the antenna with differing amplitudes, the strongest being those whose resonant frequency is close to the frequency of the excitation. This effect should be taken into account when the highest accuracy is required. But the general characteristics of the antenna are controlled by the nearest mode in frequency provided that the feed is not positioned so as to excite it only weakly. The basic analysis of this chapter will therefore assume that only the near-resonant mode exists; however some consideration will be given to methods of allowing for the presence of the other modes in Section 4.5.1.

4.1.2 Calculation of radiation fields
Some discussion of the available methods of calculation of the fields radiated by microstrip structures has already been given in Chapter 3, where it was shown that radiation from microstrip transmission line discontinuities may be treated equally from the point of view of electric current distributions or edge slot field sources, which may be represented by equivalent magnetic currents. Similarly, both methods have been applied to the calculation of the radiation of patch antennas, with comparable results being obtained. However, the method based on slot field radiation is the more straightforward since it only requires a line integration of the peripheral source whereas that based on current sources involves integration over

the whole patch area of both a physical current and an effective polarisation current. Furthermore, practical experience shows that an engineering 'feel' for a microstrip antenna is easier to obtain with the slot field approach, so this method will be relied on for the approximate analysis of this Chapter. The technique of

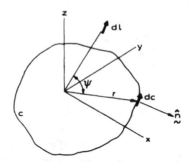

Fig. 4.7   *Calculation of radiation from patch by magnetic current method*

replacing the electric field at the edge of the microstrip circuit by a magnetic current has been discussed in Chapter 3; on this basis and taking V as the voltage at the patch edge with respect to ground the far-field electric radiation vector is given by

$$\mathbf{L} = -2 \int_c \mathbf{z} \times \mathbf{n} \, V(c) \exp\left(jk_0 r \cos \Psi\right) dc \tag{4.4}$$

where the integration range C is the closed loop around the patch effective edge, Fig. 4.7. The radiation fields due to the antenna are derived from the vector $\mathbf{L}$ by the relations

$$E_\theta = -j \, \frac{L_\phi}{2\lambda_0 R} \, \exp\left(-jk_0 R\right)$$

$$E_\phi = j \, \frac{L_\theta}{2\lambda_0 R} \, \exp\left(-jk_0 R\right) \tag{4.5}$$

and $H_\theta = -E_\phi/Z; H_\phi = E_\theta/Z$.

The power radiated P_r is given by an integration of the Poynting vector as

$$P_r = \int_0^{\pi/2} d\theta \int_0^{2\pi} d\phi \, \frac{|\mathbf{L} \times \mathbf{r}|^2}{8\lambda_0^2 Z} \, \sin \theta \tag{4.6}$$

4.1.3 Antenna bandwidth and impedance

The bandwidth of an antenna in a practical system depends upon how severe an effect the variation of the antenna characteristics with frequency has upon the over-all system performance. Although in a particular situation any of the antenna parameters may prove to be critically limiting on the antenna bandwidth, in most cases it is the antenna VSWR which limits the performance, and this is particularly true for strongly resonant devices. Even then, the tolerance set on the VSWR by the system varies, so to enable the general characterisation of a microstrip antenna, use will be made of the circuit concept of the Q-factor, Q_T. This may be defined as (Aitken, 1976) the ratio of the resonant frequency of the device and the frequency band Δf over which the power reflected is not more than one fifth (-7 dB) of that absorbed at resonance when matched to the feed line, Fig. 4.8, thus

$$Q_T = \frac{f_r}{\Delta f} \tag{4.7}$$

If a system VSWR of $< S$ is required, then the usable bandwidth of the antenna is related to the Q-factor by

$$\text{bandwidth} = \frac{100\,(S-1)}{Q_T\,\sqrt{S}}\ \%\ ; \qquad (S \geqslant 1) \tag{4.8}$$

It should be noted that this relationship is only valid whilst the antenna behaves essentially as a simple tuned RLC circuit.

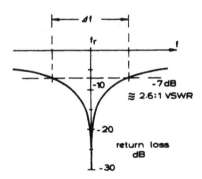

Fig. 4.8   *Return loss characteristic of resonant element*

The power absorbed by the resonant antenna includes power dissipated in the loss mechanisms as well as the power radiated into the far field, these mechanisms being conduction losses in the imperfect metal and dielectric of the substrate. The total Q-factor of the antenna may be regarded as being the combination of the Q-factors which would have been obtained if each of the power absorption mechanisms had been present alone, as stated in Chapter 3, Section 3.5; the relationship being

$$\frac{1}{Q_T} = \frac{1}{Q_R} + \frac{1}{Q_c} + \frac{1}{Q_d} \tag{4.9}$$

For efficient antenna operation the radiation Q_R should be the dominant factor as discussed in Section 3.5.

A number of formulas are available which give Q-factor as a function of device parameters, the most useful expression in the present case relates Q-factor to absorbed power and stored energy as

$$Q_\alpha = 2\pi \frac{\mathscr{E}_s f_r}{P_\alpha} \qquad (4.10)$$

where $\mathscr{E}_s$ is the total energy stored within the antenna at resonance, which is constant, although being exchanged continuously between the electric and magnetic fields. The subscript α denotes radiation, conductor or dielectric losses as appropriate using the symbols R, c and d, respectively. The energy stored in the antenna is equal to the peak energy stored by the electric and magnetic field distributions within the substrate, which leads to the formulas

$$\mathscr{E}_s = \frac{1}{2} \int_V \epsilon_r \epsilon_0 \, |E_z|^2 \, dv = \frac{1}{2} \int_V \mu_0 \, |H|^2 \, dv$$

$$= \frac{h}{2} \int_A \epsilon_r \epsilon_0 |E_z|^2 \, da = \frac{h}{2} \int_A \mu_0 \, |H|^2 \, da \qquad (4.11)$$

where the surface of integration A is the planar area of the patch. Eqn. 4.5 gives the required expression to evaluate the radiated power, and hence Q_R, and the calculation of the power dissipated by losses in the antenna will now be outlined.

If the bulk conductivity of the dielectric material is σ_d the power absorbed by the dielectric is

$$P_d = \frac{1}{2} \int_V \sigma_d \, |E_z|^2 \, dv$$

$$= \frac{h}{2} \sigma_d \int_A |E_z|^2 \, da \qquad (4.12)$$

Combining this with eqn. 4.11 gives

$$P_d = \frac{\sigma_d \mathscr{E}_s}{\epsilon_r \epsilon_0} \qquad (4.13)$$

$$Q_d = \frac{2\pi \mathscr{E}_s f_r}{P_d} = \frac{2\pi \epsilon_r \epsilon_0 f_r}{\sigma_d} \qquad (4.14)$$

Losses in dielectric materials are normally specified in terms of their loss tangent, $\tan \delta$, which is related to the bulk conductivity by $\sigma_d = 2\pi f_r \epsilon_r \epsilon_0 \tan \delta$, so that

$$Q_d = \frac{1}{\tan \delta} \qquad (4.15)$$

Thus the dielectric loss Q-factor of a microstrip antenna is independent of the dimensions of the antenna.

Finally, the conductance losses are determined by means of the relation

$$P_c = 2 \times \frac{1}{2} \int_A |I|^2 R_s da \qquad (4.16)$$

where $|I|$ is the magnitude of the current density on the top and bottom surfaces of the magnetic wall resonator model and R_s is the surface resistance of the conductor which is related (eqn. 2.11) to the bulk conductivity σ_c by

$$R_s = \sqrt{\frac{\pi f_r \mu_0}{\sigma_c}} \qquad (4.17)$$

Since $I = n \times H$, the formula becomes

$$P_c = \sqrt{\frac{\pi f_r \mu_0}{\sigma_c}} \int_A |H|^2 da$$

$$= 2 \sqrt{\frac{\pi f_r}{\mu_0 \sigma_c}} \frac{\mathcal{E}_s}{h} \qquad (4.18)$$

after combination with eqn. 4.11. The conductance loss Q-factor is thus

$$Q_c = h \sqrt{\mu_0 \pi f_r \sigma_c} \qquad (4.19)$$

and within the limitations of the magnetic wall model this is independent of the resonator geometry other than through f_r.

It should be remembered in utilising these formulas for dielectric and conductor losses that they are only as accurate as the closed-resonator model represents the true field distribution of the antenna. Analysis of microstrip transmission lines by many authors has shown that the current and electric field distributions have very strong peaks at the edges of the conductor, and this also will be the case with patch antennas. The use of the 'magnetic wall' model prevents this effect from being considered, however this should not cause too large errors in the case of a patch antenna since the contribution from the edge region to the integrals will be comparatively small. For the microstrip transmission line the edge region occupies a considerable proportion of the line width and has a correspondingly larger effect on this type of calculation. Some caution may be required however when applying the results to antennas at relatively high frequencies where the edge effects become more significant.

It is appropriate at this stage to consider the particular characteristics of some typical geometries, and this is carried out in the next two Sections of this Chapter.

4.2 Analysis of rectangular patch using modal method

4.2.1 Radiation fields
The field components of the rectangular patch elements, Fig. 4.9, are the duals of those of the well known rectangular waveguide transverse electric modes:

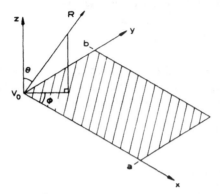

Fig. 4.9   *Co-ordinate system of rectangular patch antenna*

$$E_z = \frac{V_0}{h} \cos\left(\frac{m\pi x}{a}\right) \cos\left(\frac{n\pi y}{b}\right)$$

$$H_x = -\frac{j\omega\epsilon_0}{k_0^2} \frac{n\pi}{b} \frac{V_0}{h} \cos\left(\frac{m\pi x}{a}\right) \sin\left(\frac{n\pi y}{b}\right) \qquad (4.20)$$

$$H_y = \frac{j\omega\epsilon_0}{k_0^2} \frac{m\pi}{a} \frac{V_0}{h} \sin\left(\frac{m\pi x}{a}\right) \cos\left(\frac{n\pi y}{b}\right)$$

where V_0 is the peak voltage at the corner of the patch. The resonant frequency of the patch is given by the same formula as for cutoff in the waveguide,

$$f_r = \frac{1}{2\sqrt{\mu_0 \epsilon_0 \epsilon_r}} \sqrt{\left(\frac{m}{a}\right)^2 + \left(\frac{n}{b}\right)^2} \qquad (4.21)$$

To evaluate the radiation fields the voltage at the patch edge must be specified. This is

$$V(y) = V_0 \cos\left(\frac{n\pi y}{b}\right) \qquad\qquad x = 0; \quad 0 \leqslant y \leqslant b$$

$$V(y) = V_0 \cos\left(\frac{n\pi y}{b}\right) \cos(m\pi) \qquad x = a; \quad 0 \leqslant y \leqslant b \qquad (4.22)$$

$$V(x) = V_0 \cos\left(\frac{m\pi x}{a}\right) \qquad\qquad y = 0; \quad 0 \leqslant x \leqslant a$$

$$V(x) = V_0 \cos\left(\frac{m\pi x}{a}\right) \cos(n\pi) \qquad y = b; \quad 0 \leqslant x \leqslant a$$

Substituting these four segments in eqn. 4.4 gives

$$\mathbf{L} = 2jV_0 \{1 - \cos(m\pi) \exp(jk_0 a \cos\phi \sin\theta)\}$$

$$\{1 - \cos(n\pi) \exp(jk_0 b \sin \phi \sin \theta)\} \left[\hat{x} \; \frac{k_0 \cos \phi \sin \theta}{k_0^2 \cos^2 \phi \sin^2 \theta - \dfrac{m^2 \pi^2}{a^2}} \right.$$

$$\left. -\hat{y} \; \frac{k_0 \sin \phi \sin \theta}{k_0^2 \sin^2 \phi \sin^2 \theta - \dfrac{n^2 \pi^2}{b^2}} \right] \tag{4.23}$$

where $\hat{x}$ and $\hat{y}$ are unit vectors.

Converting these to spherical components and substituting in eqn. 4.5 gives the general expressions for the field components as

$$E_\theta = -\frac{\exp(-jk_0 R)}{2\lambda_0 R} \; 2V_0 \{1 - \cos(m\pi) \exp(jk_0 a \cos \phi \sin \theta)\}$$

$$\{1 - \cos(n\pi) \exp(jk_0 b \sin \phi \sin \theta)\} k_0 \sin \theta \cos \phi \sin \phi$$

$$\left\{ \frac{1}{k_0^2 \cos^2 \phi \sin^2 \theta - \dfrac{m^2 \pi^2}{a^2}} + \frac{1}{k_0^2 \sin^2 \phi \sin^2 \theta - \dfrac{n^2 \pi^2}{b^2}} \right\} \tag{4.24a}$$

$$E_\phi = \frac{\exp(-jk_0 R)}{2\lambda_0 R} \; 2V_0 \{1 - \cos(m\pi) \exp(jk_0 a \cos\phi \sin\theta)\}$$

$$\{1 - \cos(n\pi) \exp(jk_0 b \sin \phi \sin \theta)\} k_0 \sin \theta \cos \theta$$

$$\left\{ \frac{\sin^2 \phi}{k_0^2 \sin^2 \phi \sin^2 \theta - \dfrac{n^2 \pi^2}{b^2}} - \frac{\cos^2 \phi}{k_0^2 \cos^2 \phi \sin^2 \theta - \dfrac{m^2 \pi^2}{a^2}} \right\} \tag{4.24b}$$

The case of most interest is the lowest order mode when $m = 0$ and $n = 1$, and this leads to the following expressions for the fields radiated:

$$E_\theta = -j \frac{\exp(-jk_0 R)}{2\lambda_0 R} \; 8V_0 \exp\left(j \frac{k_0 a}{2} \cos \phi \sin \theta\right) \exp\left(j \frac{k_0 b}{2} \sin \phi \sin \theta\right)$$

$$\sin\left(\frac{k_0 a}{2} \cos \phi \sin \theta\right) \cos\left(\frac{k_0 b}{2} \sin \phi \sin \theta\right) k_0 \sin \phi \cos \phi \sin \theta$$

$$\left\{ \frac{1}{k_0^2 \cos^2 \phi \sin^2 \theta} - \frac{1}{k_0^2 \sin^2 \phi \sin^2 \theta - \dfrac{\pi^2}{b^2}} \right\} \tag{4.25a}$$

$$E_\phi = -j \frac{\exp(-jk_0 R)}{2\lambda_0 R} \; 8V_0 \exp\left(j \frac{k_0 a}{2} \cos \phi \sin \theta\right) \exp\left(j \frac{k_0 b}{2} \sin \phi \sin \theta\right)$$

$$\sin\left(\frac{k_0a}{2}\cos\phi\sin\theta\right)\cos\left(\frac{k_0b}{2}\sin\phi\sin\theta\right)k_0\sin\theta\cos\theta$$

$$\left\{\frac{1}{k_0^2\sin^2\theta}-\frac{\sin^2\phi}{k_0^2\sin^2\phi\sin^2\theta-\pi^2/b^2}\right\}\tag{4.25b}$$

The second term in the expression for each component of the radiated field is zero in the two principal ϕ planes. This is due to the fact that these terms originate from the sides of the patch where the magnetic current distribution is antisymmetric about both axes of the patch. The reduced expressions for the principal planes are

$$E_\theta\ (\phi=90^\circ)=-j\frac{\exp(-jk_0R)}{2\lambda_0R}4V_0a\exp\left(j\frac{k_0b}{2}\sin\theta\right)\cos\left(\frac{k_0b}{2}\sin\theta\right)$$

$$E_\phi\ (\phi=90^\circ)=0\tag{4.26a}$$

$$E_\theta\ (\phi=0^\circ)\ \ =0$$

$$E_\phi\ (\phi=0^\circ)\ \ =-j\frac{\exp(-jk_0R)}{2\lambda_0R}4V_0a\exp\left(j\frac{k_0a}{2}\sin\theta\right)\cos\theta\frac{\sin\left(\frac{k_0a}{2}\sin\theta\right)}{\frac{k_0a}{2}\sin\theta}$$

$$\tag{4.26b}$$

Thus one principal plane pattern corresponds to an interferometer antenna, and the other to the product of a uniform distribution pattern and a cosine current element pattern as previously noted in Section 3.5, Chapter 3. The secondary terms reach their maximum values in the $\phi=45^\circ$ planes, contributing radiation of $\sim-16\,\mathrm{dB}$ relative to the signal at $\theta=0^\circ$ near the ground plane. This comparatively weak level suggests that these terms may be neglected in calculating the radiation patterns of a single patch antenna, particularly when it is noted that their main contribution lies in directions which will be strongly influenced by other factors such as diffraction from the edges of the physically limited ground-plane. It should be remembered,

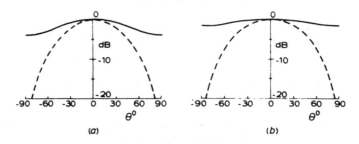

(a)   (b)

Fig. 4.10   Theoretical radiation patterns of square patch antenna in $\phi=45^\circ/225^\circ$ plane
a Dominant terms of eqn. 4.25 only
b All terms of eqn. 4.25
———— $|E_\theta|$
– – – $|E_\phi|$
$\epsilon_r=2\cdot0$

however, that in calculating the radiation power of the antenna, cross-producting of the dominant and secondary terms occurs which may lead to conductance errors of up to 10% when these terms are neglected; also they may be of significance when considering cross-polarised radiation from arrays of patches (Chapter 5). Fig. 4.10 shows an example of the differences between the radiation patterns obtained when the secondary terms are not taken into account.

4.2.2 Radiation Q-factor

Evaluation of the power radiated by the antenna operated in a general mode requires substitution of the formula for L given by eqn. 4.23 into eqn. 4.6. Because of the complexity of the general expression for L, the integral is best evaluated by numerical methods for particular cases.

Having calculated the radiated power, the remaining task is to evaluate the stored energy in the patch. Substituting the expression for E_z, eqn. 4.20 into eqn. 4.11 gives

$$\mathcal{E}_s = \frac{\epsilon_r \epsilon_0 V_0^2}{2h} \int_0^a \int_0^b \cos^2\left(\frac{m\pi x}{a}\right) \cos^2\left(\frac{n\pi y}{b}\right) dxdy$$

$$= \frac{\epsilon_r \epsilon_0 V_0^2 ab}{2h} \delta_m \delta_n \tag{4.27}$$

where

$$\delta_l = 1, \quad l = 0$$
$$= \tfrac{1}{2}, \quad l = 1, 2, 3 \dots$$

Plots of Q_R derived using eqn. 4.10 for the ($m = 0, n = 1$) mode are given in Fig. 4.11 as a function of a/b for various values of ϵ_r. The overall Q-factor of the antenna should, of course, include the effects of the losses in conductor and dielectric as in Section 4.1.3.

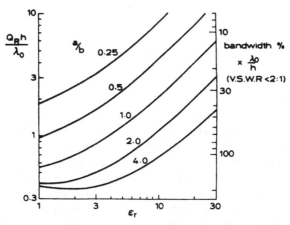

Fig. 4.11   *Theoretical Q_R and bandwidth of rectangular patch antennas as function of parameters*

4.2.3. Square antenna

If the rectangular patch is reduced to a square, $a = b$, the mode sets with $m^2 + n^2$ = constant are resonant at the same frequency. In the case of the $(m = 0, n = 1)$ and $(m = 1, n = 0)$ mode pair, the resultant polar diagrams are identical, but rotated in the ϕ-plane by $90°$ throughout. The radiated fields in the $\theta = 0°$ direction for these two modes are then orthogonally polarised, so that if the modes are excited with equal amplitude but in $90°$ phase difference, circular polarisation is radiated on boresight. Since the E- and H-plane patterns of each mode differ, the circular polarisation will deteriorate as the observation angle $\theta°$ is increased with linear polarisation occuring at $\theta = 90°$ as $E_\phi \to 0$. The ellipticity is quite acceptable over the 3 dB — beamwidth of the wanted circular polarisation, radiation patterns for the principal planes being shown in Fig. 4.12 for $\epsilon_r = 2$. Since the patch is not rotationally symmetric, the pattern varies slightly as a function of ϕ with a maximum of typically ± 2dB at $\theta = 90°$.

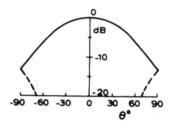

Fig. 4.12 *Theoretical radiation patterns of circular polarised square patch antenna in $\phi = 0°/180°$ plane*
          ──── *wanted hand of c.p.*
          ─ ─ ─ *orthogonal hand of c.p.*
          $\epsilon_r = 2\cdot 0$

4.3 Analysis of circular patch using modal method

4.3.1. Radiation fields

The circular patch antenna is a special case of the elliptical patch, corresponding to the specialisation of the rectangular patch to the square. The electromagnetic analysis of structures with elliptical boundaries is complicated since it requires use of Mathieu Functions. The use of elliptical resonators in microstrip circulators has been discussed by Irish (1971) and Kretzshmar (1972); Long and Shen (1981) have analysed elliptical shaped microstrip antennas. The following discussion is restricted to circular antennas, in which case the use of a cylindrical coordinate system is appropriate, Fig. 4.13, leading to the use of Bessel Functions in the field equations, as follows:

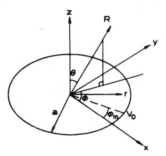

Fig. 4.13   *Co-ordinate system of circular patch antenna*

$$E_z = \frac{V_0}{h} \frac{J_n(\sqrt{\epsilon_r}\, k_0 r)}{J_n(\sqrt{\epsilon_r}\, k_0 a)} \cos\left(n(\phi - \phi_m)\right)$$

$$H_r = -j\frac{V_0}{h} \frac{\omega n \epsilon_0}{k_0^2 r} \frac{J_n(\sqrt{\epsilon_r}\, k_0 r)}{J_n(\sqrt{\epsilon_r}\, k_0 a)} \sin\left(n(\phi - \phi_m)\right) \qquad (4.28)$$

$$H_\phi = -j\frac{V_0}{h} \frac{\omega\sqrt{\epsilon_r}\,\epsilon_0}{k_0} \frac{J'_n(\sqrt{\epsilon_r}\, k_0 r)}{J_n(\sqrt{\epsilon_r}\, k_0 a)} \cos\left(n(\phi - \phi_m)\right)$$

where V_0 is the peak voltage at the edge of the patch at an angle $\phi = \phi_m$. The resonant frequency of the patch is defined by the characteristic equation

$$J'_n(\sqrt{\epsilon_r}\, k_0 a) = 0 \qquad (4.29)$$

No closed-form expression exists for the roots x of $J'_n(x)$ and the first few are listed in order of magnitude in Table 4.1 for convenience.

Table 4.1   *Roots of $J'_n(x) = 0$*

x	n	x	n
1·841	1	5·331	1
3·054	2	6·416	5
3·832	0	6·706	2
4·201	3	7·016	0
5·317	4	7·501	6

One difference between the field formulas given in eqn. 4.20 for the rectangular patch and those of eqn. 4.28 is that whereas the integer values of m, n for the rectangular patch give rise to discrete modes, the range of ϕ_m for the circular patch is continuous. This is simply due to the lack of preferred axes in the case of the circular patch.

The voltage at the patch edge which acts as the radiation source is

$$V(\phi) = V_0 \cos(n(\phi - \phi_m)) \tag{4.30}$$

Substituting in eqn. 4.4 gives

$$L = 2j\pi a V_0 \exp\left(jn\frac{\pi}{2}\right) \{\hat{r} \sin(n(\phi - \phi_m)) \sin\theta [J_{n+1}(z) + J_{n-1}(z)]$$

$$+ \hat{\theta} \sin(n(\phi - \phi_m)) \cos\theta [J_{n+1}(z) + J_{n-1}(z)]$$

$$+ \hat{\phi} \cos(n(\phi - \phi_m))[J_{n+1}(z) - J_{n-1}(z)] \} \tag{4.31}$$

where

$$z = k_0 a \sin\theta \text{ and } \hat{r}, \hat{\theta}, \hat{\phi} \text{ are unit vectors}$$

and from eqn. 4.5 the radiated electric field components are

$$E_\theta = \frac{\exp(-jk_0 R)}{\lambda_0 R} \pi a V_0 e^{j(n\pi/2)} \cos(n(\phi - \phi_n))[J_{n+1}(z) - J_{n-1}(z)] \tag{4.32a}$$

$$E_\phi = -\frac{\exp(-jk_0 R)}{\lambda_0 R} \pi a V_0 e^{j(n\pi/2)} \sin(n(\phi - \phi_m)) \cos\theta [J_{n+1}(z)$$

$$+ J_{n-1}(z)] \tag{4.32b}$$

As with the rectangular case, the lowest order mode $n = 1$ is of particular interest, so

$$E_\theta = -j\frac{\exp(-jk_0 R)}{\lambda_0 R} \pi a V_0 \cos(\phi - \phi_m)[J_0(k_0 a \sin\theta) - J_2(k_0 a \sin\theta)] \tag{4.33a}$$

$$E_\phi = -j\frac{\exp(-jk_0 R)}{\lambda_0 R} \pi a V_0 \sin(\phi - \phi_m) \cos\theta [J_0(k_0 a \sin\theta)$$

$$+ J_2(k_0 a \sin\theta)] \tag{4.33b}$$

In this case, $E_\phi = 0$ in the $\phi = \phi_m$ plane and $E_\theta = 0$ in the $\phi = \phi_m + 90°$ plane. These are the principal planes for this antenna, and the polar diagrams are shown in Fig. 4.14. It should be noted that there is an infinite set of modes for $n = 1$, defined by the roots of eqn. 4.29, and the results shown are for the first mode $\sqrt{\epsilon_r}$ $k_0 a = 1 \cdot 841$. Increasing the mode number leads to polar diagrams with narrowed beamwidths but high sidelobe levels, which is a characteristic of ring sources.

4.3.2 Radiation Q-factor
The radiated power of the circular antenna is evaluated by substituting L as given in eqn. 4.31 into eqn. 4.6. This leads to

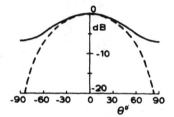

Fig. 4.14 *Theoretical radiation patterns of circular patch antenna in principal planes*
——— $|E_\theta|$ in $\phi = 0°/180°$ plane
– – – $|E_\phi|$ in $\phi = 90°/270°$ plane
$\epsilon_r = 2\cdot0$

$$P_r = \int_0^{\pi/2} d\theta \int_0^{2\pi} d\phi \, \frac{\pi^2 a^2 V_0^2 \sin\theta}{2\lambda_0^2 Z} \{\cos^2(n(\phi - \phi_m))[J_{n+1}(z) - J_{n-1}(z)]^2$$

$$+ \sin^2(n(\phi - \phi_m)) \cos^2\theta \, [J_{n+1}(z) + J_{n-1}(z)]^2\}$$

$$= \frac{\pi^3 a^2 V_0^2}{\lambda_0^2 Z} \delta_n \int_0^{\pi/2} \left[2J_{n-1}^2(z) - \frac{4n^2}{k_0^2 a^2} J_n^2(z) + 2J_{n+1}^2(z)\right] \sin\theta \, d\theta \quad (4.34)$$

where

$$\delta_n = 1, \quad n = 0$$
$$= \tfrac{1}{2}, \quad n \neq 0$$

This may be evaluated in series form using the identities

$$J_m(z)J_n(z) = \sum_{k=0}^{\infty} \frac{(-1)^k \left(\dfrac{z}{2}\right)^{m+n+2k} (m + n + 2k)!}{(m + n + k)!(m + k)!(n + k)!k!}$$

$$\int_0^{\pi/2} \sin^{2q+1}\theta \, d\theta = \frac{2^{2q}(q!)^2}{(2q + 1)!}$$

giving

$$P_r = \frac{\delta_n \pi^3 a^2 V_0^2}{\lambda_0^2 Z} \left[\frac{4n(n + 1)(k_0 a)^{2n-2}}{(2n + 1)!} - \frac{8n(n + 1)(n + 3)(k_0 a)^{2n}}{(2n + 3)!} \right.$$

$$+ \sum_{k=0}^{\infty} \frac{(-1)^k 2(k_0 a)^{2n+2k+2}}{(2n + k + 2)!(2n + 2k + 5)(2n + 2k + 3)(k + 2)!} \times$$

$$[(2n + k + 1)(2n + k + 2)(2n + 2k + 5) - 2n^2(2n + 2k + 3)$$

$$\left. + (k + 1)(k + 2)(2n + 2k + 5)] \right] \quad (4.35)$$

For the $n = 1$ mode, this becomes

$$P_r = \frac{\pi^3 a^2 V_0^2}{2\lambda_0^2 Z}\left[\frac{4}{3} - \frac{8}{15}\left(\frac{1\cdot841}{\epsilon_r}\right) + \frac{11}{105}\left(\frac{1\cdot841}{\epsilon_r}\right)^2 \cdots\right] \qquad (4.36)$$

The total power stored in the resonator is given by

$$\mathscr{E}_s = \frac{\epsilon_0 \epsilon_r h}{2}\frac{V_0^2}{h^2}\int_0^a\int_0^{2\pi}\frac{J_n^2(\sqrt{\epsilon_r}k_0 r)}{J_n^2(\sqrt{\epsilon_r}k_0 a)}\cos^2(n(\phi - \phi_m))\,r\,dr\,d\phi$$

$$= \frac{\pi V_0^2 \delta_n}{2h\omega^2\mu_0}\left[\epsilon_r k_0^2 a^2 - n^2\right] \qquad (4.37)$$

Plots of Q_R derived using eqn 4.10 for the lowest $n = 1$ mode are shown in Fig. 4.15 for various values of ϵ_r, showing very similar results to those obtained for the square patch given in Fig. 4.11. Again, the overall Q-factor of the antenna requires allowances for the losses of the materials, given for instance by eqns. 4.15 and 4.19.

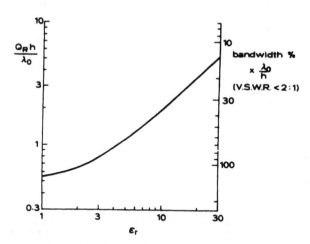

Fig. 4.15   Theoretical Q_R and bandwidth of circular patch antenna as function of substrate permittivity

4.3.3 Circular polarisation

Although the ϕ_m can take a continuous range of values, orthogonality of the radiated fields is only obtained for pairs of modes with the values ϕ'_m and $90° + \phi'_m$. In this case, separate excitation is possible and with phase quadrature between the modes, circular polarisation is radiated on boresight. Variation of the circular polarisation with angle is shown in the principal plane patterns of Fig. 4.16 for $\epsilon_r = 2$, with similar results again being obtained to that of the square patch antenna Fig. 4.12 but in this case the patterns are independent of ϕ since the patch is circularly symmetric.

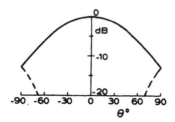

Fig. 4.16 *Theoretical radiation patterns of circularly polarised circular patch antenna*
——— wanted hand of c.p.
— — — orthogonal hand of c.p.
$\epsilon_r = 2 \cdot 0$
Patterns are valid for all ϕ-plane cuts due to the rotational symmetry of the antenna

4.4 Input impedance analysis

4.4.1 Network model for (m = 0, n = 1) mode rectangular patch

Analysis of the rectangular patch antenna is simplified in the case of a patch operating in the $(0, n)$ mode by regarding the patch as an $n\lambda_m/2$ wavelength microstrip transmission-line resonator, where the radiation is regarded as being due to the two ends of the line acting as open-circuit line radiators, as discussed in Chapter 3. The analysis has to take account of the effects of mutual coupling between the end radiators in the present configuration, which is of particular significance for the half-wavelength resonator, and so the network model of Fig. 4.17 is adopted. The feed point P may represent either a microstrip transmission line coupled to the edge of the patch antenna, or a coaxial line feed passing through the substrate. The susceptance B is representative of the effects of the fringing fields at the end of the line, and could alternatively be represented as an effective line extension Δl as in Chapter 2. The self-conductance G_r has been derived in Chapter 3 as

$$G_r = \frac{1}{120\pi^2} F_2\left(\frac{2\pi}{\lambda_0} w_e\right) \tag{4.38}$$

where

$$F_2(x) = xSi(x) - 2\sin^2(x/2) - 1 + \sin(x)/x$$

This formula for G_r may be replaced with good accuracy by the following three relations, depending on the value of $w_e/\lambda_0 :-$

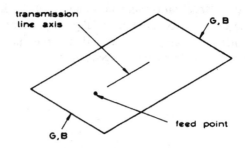

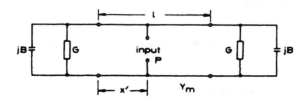

Fig. 4.17   *Transmission line network model of rectangular patch antenna*

$$G_r = w_e^2/90\lambda_0^2, \qquad\qquad w_e < 0.35\lambda_0$$

$$G_r = w_e/120\lambda_0 - 1/60\pi^2, \qquad 0.35\lambda_0 \leqslant w_e \leqslant 2\lambda_0 \qquad (4.39)$$

$$G_r = w_e/120\lambda_0, \qquad\qquad 2\lambda_0 < w_e$$

The mutual conductance G_m between the ends may be determined by integrating the interference component of the far-field radiation patterns of two magnetic current sources of length w_e spaced by a distance l perpendicular to their length (Derneryd, 1978).

$$G_m = \frac{1}{120\pi^2} \int_0^\pi \frac{\sin^2\left(\dfrac{\pi w_e}{\lambda_0} \cos\theta\right)}{\cos^2\theta} \sin^3(\theta) J_0\left(\frac{2\pi l \sin\theta}{\lambda_0}\right) d\theta \qquad (4.40)$$

In addition to the mutual conductance, there will be a mutual susceptance, component which could be evaluated by extending the range of integration of the radiation interference pattern into the imaginary angle region. In the case of a patch antenna, however, the two sources are inherently phased at either $0°$ or $180°$, so that only the mutual conductance term will contribute to the total radiation conductance, the mutual susceptance causing a slight detuning of the resonant frequency. The expression for mutual susceptance is more complicated than that for mutual conductance, involving double integrals of two terms, and will not be considered here. The mutual conductance is shown in Fig. 4.18 as a function of spacing for different values of w_e.

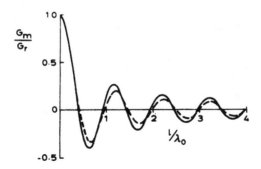

Fig. 4.18 *Mutual conductance between radiating ends of rectangular patch antenna*
After Derneryd (1978)
——— $w = \lambda_0$
– – – $w = 0.3\lambda_0$

If the total admittance at each radiating end of the line is $G + jB$, which includes both self- and mutual-impedances, and the microstrip transmission line admittance is Y_m, then at resonance the input impedance at the feed point is (Derneryd (1977))

$$Z_{in}(x') = \frac{1}{2G} \left\{ \cos^2(\beta x') + \frac{G^2 + B^2}{Y_m^2} \sin^2(\beta x') - \frac{B}{Y_m} \sin(2\beta x') \right\}$$

$$\sim \frac{1}{2G} \cos^2(\beta x') \tag{4.41}$$

since $G, B \ll Y_m$ for typical microstrip structures. Splitting the conductance into the self- and mutual-components gives

$$Z_{in}(x') = \frac{1}{G_r + G_m \cos(n\pi)} \cos^2(\beta x') \tag{4.42}$$

for the $(0, n)$ mode. This expression gives a comparatively simple method of evaluating the resonant input impedance as a function of position of the feed, although it should be noted that it has been obtained by neglecting the effects of losses in the conductor and dielectric.

The antenna Q-factor may be calculated by evaluating the frequencies for which the reactive component of the input impedance is equal to the input resistance. However, the full expressions for the impedance, or admittance, of the antenna are very complicated for frequencies away from resonance. A simpler method of estimating the Q due to radiation loading is the Fabry-Perot formula eqn. 3.29 for the Q-factor of a $q\lambda_m/2$ long transmission line with equal reflection coefficients Γ at each end

$$Q = \frac{q\pi |\Gamma|}{1 - |\Gamma|^2} \qquad (4.43a)$$

$$q = 1, 2, \ldots$$

The reflection coefficient magnitude is given by

$$|\Gamma| = \sqrt{\frac{(Y_m - G)^2 + B^2}{(Y_m + G)^2 + B^2}} \qquad (4.43b)$$

for a terminal admittance $(G + jB)$

If it is assumed that B is small and $|\Gamma| \sim 1$ eqn. 4.43a may be approximated by

$$Q_R = \frac{\pi Y_m}{4G_r} \qquad (4.44)$$

Substituting the simplified formulas for G_r from eqn. 4.39 and using eqn. 2.16 leads to some useful expressions for estimating the radiation Q of rectangular patch antennas. Neglecting mutual coupling between edges we have

$$Q_R = \frac{3}{16} \frac{\epsilon_r}{ZY_m} \frac{\lambda_0^2}{h^2}, \qquad w_e < 0.35\lambda_0 \qquad (4.45a)$$

$$Q_R = \frac{\pi Y_m}{\dfrac{4\pi Y_m}{\sqrt{\epsilon_r}} \dfrac{h}{\lambda_0} - \dfrac{1}{15\pi^2}}, \qquad 0.35\lambda_0 \leqslant w_e \leqslant 2\lambda_0 \qquad (4.+5b)$$

$$Q_R = \frac{\sqrt{\epsilon_r}}{4} \frac{\lambda_0}{h}, \qquad 2\lambda_0 < w_e \qquad (4.45c)$$

In the case of a square patch, the equivalent width is equal to a half-wavelength of the microstrip, so if allowance is made for the effects of field fringing (see Section 4.5.2)

$$w_e = \frac{\lambda_0}{2\sqrt{\epsilon_r}} \qquad (4.46a)$$

and

$$Y_m = \frac{\lambda_0}{2Zh} \qquad (4.46b)$$

Substituting in eqn. 4.45a and b gives

$$Q_R = \frac{3}{8} \epsilon_r \frac{\lambda_0}{h}, \qquad \epsilon_r > 2 \qquad (4.47a)$$

$$Q_R = \frac{\pi^2 \sqrt{\epsilon_r}}{4\pi^2 - 16\sqrt{\epsilon_r}} \frac{\lambda_0}{h}, \qquad \epsilon_r < 2 \qquad (4.47b)$$

In fact, the difference between these formulas for $\epsilon_r < 2$ is $< 10\%$, which in view of the approximations made in the derivation suggests that eqn. 4.47a alone is adequate for indicating the Q_R of a square microstrip patch antenna. This is a particularly useful formula for $\epsilon_r < 4$ since the square patch is a commonly used device, and also it may be assumed that it would give a reasonable figure for a circular patch of equivalent area. An estimate of the bandwidth of a square or circular patch for a VSWR S can be made by combining eqn 4.47a with eqn 4.8 to give

$$\text{bandwidth} = \frac{100(S-1)}{\sqrt{S}} \frac{8}{3\epsilon_r} \frac{h}{\lambda_0} \% \tag{4.48}$$

It is worth noting that if the dimensions of the patch are less than $10h$, the relative permitivity ϵ_r should be replaced in the preceding formulas by the effective permitivity ϵ_e. This is a further allowance for the effects of field fringing, discussed in more detail in Section 4.5.2.

4.4.2 General patch shapes

The preceding Section gives a method of calculating the input impedance of a patch which is only directly applicable to the case of a rectangular patch operating in the $(0, 1)$ mode. The concept of the nearly open-circuit transmission-line segment may be used to analyse a number of other forms; the circular patch, for example, could be regarded as a radial transmission line loaded with a small conductance at its periphery, but this type of approach does not give a physical insight into the operation of the more complex structures. It is more instructive in general to return to the ideas of Section 4.1 and regard the patches as lossy resonators, the input conductance at resonance being calculated from the power radiated and dissipated by the relation

$$G_{in} = G_r + G_d + G_c$$

$$= \frac{2P_r}{|V_f|^2} + \frac{2P_d}{|V_f|^2} + \frac{2P_c}{|V_f|^2} \tag{4.49}$$

where V_f is the resonant mode voltage at the feed point. Because the voltage between the plates varies as a function of feed position, the conductance seen by the feed line may be controlled so as to match to any desired admittance by correctly positioning the feed point. Table 4.2 gives the radiation conductance of various patch antennas referred to the voltage maximum at the periphery.

In general the admittance of a patch antenna referred to the edge is much smaller than the characteristic conductance of feed lines conventionally used in practice; coaxial cables in particular commonly have a characteristic conductance of 0·02 S. A microstrip antenna may have a markedly larger conductance than those given in Table 4.2 if the dissipation losses associated with it are comparatively large, a situation which is likely for antennas with small thickness in terms of wavelengths. Although this may make matching the antenna to a feed system easier, it must be

Table 4.2 *Radiation conductance of patches referred to edge voltage maxima, $\epsilon_r = 2.5$, fundamental modes*

Patch type	G_r mS
Disc	2·20
Square	2·68
2 × 1 rectangle	8·96

remembered that the presence of large loss conductances means the overall efficiency of the antenna will be reduced since

$$\text{efficiency } \eta = \frac{G_r}{G_r + G_d + G_c} \times 100\% \tag{4.50}$$

The susceptance B of a resonant structure is related to the difference between the peak stored electric and magnetic energy within the structure,

$$B = 2\omega\, (\&_e - \&_m)/|V_f|^2 \tag{4.51}$$

When suitable expressions are substituted for $\&_e$ and $\&_m$ in this formula, the result is a susceptance which near resonance behaves in the same way as a parallel tuned LC circuit as a function of frequency; this allows a simplified model. As an example, the stored energies in a rectangular patch with peak voltage V_0 as a function of frequency are

$$
\&_e = \int_0^a \int_0^b \frac{1}{2}\, \epsilon_0 \epsilon_r\, \frac{|V_0|^2}{h^2}\, \cos^2\left(\frac{m\pi x}{a}\right) \cos^2\left(\frac{n\pi y}{b}\right) h\,dx\,dy
$$

$$
= \frac{1}{8}\, \epsilon_0 \epsilon_r\, \frac{|V_0|^2}{h}\, ab \tag{4.52}
$$

$$
\&_m = \int_0^a \int_0^b \frac{1}{2}\, \mu_0\, \frac{\omega^2 \epsilon_0^2}{k_0^4}\, \frac{|V_0|^2}{h^2} \left\{ \frac{n^2\pi^2}{b^2} \cos^2\left(\frac{m\pi x}{a}\right) \sin^2\left(\frac{n\pi y}{b}\right) + \right.
$$

$$
\left. \frac{m^2\pi^2}{a^2} \sin^2\left(\frac{m\pi x}{a}\right) \cos^2\left(\frac{n\pi y}{b}\right) \right\} h\,dx\,dy
$$

$$
= \frac{1}{8}\, \mu_0\, \frac{\omega^2 \epsilon_0^2}{k_0^4}\, \frac{|V_0|^2}{h}\, ab \left\{ \frac{n^2\pi^2}{b^2} + \frac{m^2\pi^2}{a^2} \right\} \tag{4.53}
$$

giving

$$
B = 2\omega \left(\epsilon_r - \frac{\omega^2 \mu_0 \epsilon_0}{k_0^4} \left\{ \frac{n^2\pi^2}{b^2} + \frac{m^2\pi^2}{a^2} \right\} \right) \frac{\epsilon_0 ab}{8h}\, \frac{|V_0|^2}{|V_f|^2}
$$

$$= \left[\frac{\omega\epsilon_0\epsilon_r ab}{4h} - \frac{ab}{\omega\mu_0 4h} \left\{ \frac{n^2\pi^2}{b^2} + \frac{m^2\pi^2}{a^2} \right\} \right] \frac{|V_0|^2}{|V_f|^2} \tag{4.54}$$

which is of the form

$$B = \left\{ \omega C - \frac{1}{\omega L} \right\}$$

as found with the parallel LC circuit. The conductance G_{in} associated with the patch will be essentially constant over the frequency range near resonance so the total admittance will be expected to have the locus of a circle on a Smith Chart with its centre on the real axis. In practice the centre is generally found to be displaced slightly to the inductive side due to the existence of other modes excited by the feed as discussed in Section 4.5.1.

An alternative method of predicting the susceptance behaviour with frequency is to use the fact that for most microstrip structures the frequency band over which the susceptance is comparable to the conductance is very small compared to the resonant frequency. The expression for susceptance using the parallel LC model may then be written as

$$B = \left\{ (\omega_r + \Delta\omega)\, C - \frac{1}{(\omega_r + \Delta\omega)L} \right\}$$

$$= \frac{2\Delta\omega}{\omega_r^2 L}$$

$$B = \frac{2\Delta\omega G_{in}}{\omega_r Q_T} \tag{4.55}$$

since $Q_T = \omega_r L G_{in}$. Thus the behaviour of the susceptance is approximately linear near resonance for narrowband microstrip antennas, and may be calculated in terms of the conductance at resonance and the Q-factor.

4.5 Refinements to theoretical models

The calculations in the previous Sections were based on the simple model of the antenna as a magnetic wall resonator having only one mode present. Although this approach gives acceptable results in many cases, refinements are required to give more accurate results. These arise from two essentially independent characteristics, the first being the necessity to make some allowances for the effects of fringing fields in terms of the location of magnetic walls, radiation sources and effective substrate permittivity. The second extension to the theory required is to note that the assumption that only one mode is present will not in general be true. The fields excited by a feed will be dominated by a particular mode near its resonant frequency, but other modes will also be present and they should be accounted for if good accuracy is required in the radiation or circuit parameters. This type of effect will be more important at frequencies away from resonance or for feed positions which do not couple strongly to the dominant mode.

A further feature of practical microstrip antennas not dealt with by the simple model is the effect of the substrate on fields away from the patch boundary, which may couple surface waves. This is very difficult to incorporate by modifying a simple model, since surface-wave analysis requires consideration of the boundary conditions at the air/dielectric interface. The more complex analyses discussed in Chapter 9 are then necessary, and some consideration of the influence of surface-wave generation will be found there.

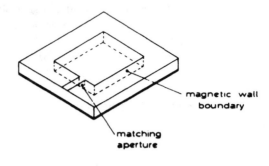

(a)

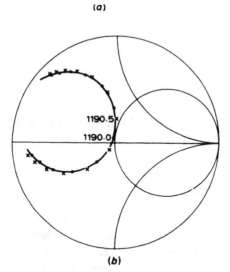

(b)

Fig. 4.19 *Analysis of patch antennas by complete modal field description*

 a Rectangular patch with surface microstrip line feed — aperture field matching methods

 b Theoretical impedance using modified permittivity method for rectangular patch with coaxial line feed

 —●— experimental results

 x calculated points, increments of 5 MHz

4.5.1 Complete modal field description

This problem has been studied by Lo *et al.* (1979) and Richards *et al.* (1979), who consider in particular a rectangular patch antenna fed at one edge by a microstrip feed line (Fig 4.19(a)) although the methods used are applicable to any geometry in principle as they point out. This model thus has an aperture feed incorporated in it which may be treated either by matching the tangential magnetic fields of the cavity modes to that of the line, or alternatively, the magnetic fields of the feed line may be replaced by a driving current source distribution over the feed aperture which is then used as a current source in the inhomogeneous Helmholtz equation. Either method may also be adapted to analyse the case of a probe feed at an arbitrary point within the patch area.

The first technique, field matching, is used when the fields are expressed as a series of waveguide modes of E_z, as in the rectangular patch case

$$E_z = \sum_m A_m \cos\left(\frac{m\pi x}{a}\right) \cos(\beta_m(y-b)) \tag{4.56a}$$

$$H_x = \sum_m \frac{A_m}{j\omega\mu_0} \beta_m \cos\left(\frac{m\pi x}{a}\right) \sin(\beta_m(y-b)) \tag{4.56b}$$

$$H_y = \sum_m -\frac{A_m}{j\omega\mu_0} \frac{m\pi}{a} \sin\left(\frac{m\pi x}{a}\right) \cos(\beta_m(y-b)) \tag{4.56c}$$

where

$$\beta_m = \sqrt{\epsilon_r k_0^2 - \frac{m^2\pi^2}{a^2}}$$

and each mode thus satisfies the homogeneous Helmholtz equation

$$(\nabla^2 + \epsilon_r k_0^2)E_z = 0$$

and the boundary condition of zero magnetic field parallel to the three sides of the patch.

A Fourier decomposition of the applied H-field, is carried out to identify the A_m coefficients with those of the modal H-field series.

The current source method is used when the fields are expressed as a series of cavity modes of E_z, i.e.

$$E_z = \sum_m \sum_n \frac{B_{mn}}{\epsilon_r k_0^2 - k_{mn}^2} \cos\left(\frac{m\pi x}{a}\right) \cos\left(\frac{n\pi y}{b}\right) \tag{4.57a}$$

$$H = -\frac{j}{\omega\mu_0} \hat{z} \times \nabla E_z \tag{4.57b}$$

where

$$k_{mn}^2 = \frac{m^2\pi^2}{a^2} + \frac{n^2\pi^2}{b^2} \quad \text{and } \hat{z} \text{ is a unit vector}$$

In this case the B_{mn} are determined by substituting E_z into the inhomogeneous Helmholtz equation

$$(\nabla^2 + \epsilon_r k_0^2)\, \mathbf{E} = j\omega\mu_0\, \mathbf{J} - \frac{1}{j\omega\epsilon_0\epsilon_r}\, \nabla\nabla\cdot\mathbf{J} \tag{4.58}$$

since $J_x = J_y = 0$ and $\partial/\partial z$ gives zero for the thin substrates, this reduces to

$$(\nabla^2 + \epsilon_r k_0^2)\, E_z = j\omega\mu_0\, J_z \tag{4.59}$$

Thus

$$\sum_m \sum_n B_{mn} \cos\left(\frac{m\pi x}{a}\right) \cos\left(\frac{n\pi y}{b}\right) = j\omega\mu_0 J_z \tag{4.60}$$

and again, Fourier decomposition may be used at the boundary to determine B_{mn} from the coefficients of a series representation of J_z.

Lo *et al.* (1979) make the point that the waveguide mode expansion of eqn. 4.56 has the advantage that it is more rapidly convergent in practice, but the cavity mode expansion of eqn. 4.57 gives greater physical insight, particularly near resonant frequencies. It is also worth bearing in mind that the radiation pattern of the antenna may either be calculated by evaluation of the total field around the periphery of the patch and using eqns. 4.4 and 4.6 or by calculating the radiation field of each mode individually with eqns. 4.4 and 4.6 and then summing the modes to determine the total radiation. If the latter method is chosen similar advantages in physical insight will occur since the cavity mode expansion will lead to patterns one of which will be very close to the total antenna pattern.

Results of the theoretical impedance locii based on the cavity mode expansions were compared with experiment by Lo *et al.* (1979); there were discrepancies in both frequency, due to the fringing field effects as discussed in the following Section, and in some cases in the shape of the locii. Richards *et al.* (1979) discuss possible reasons for the latter errors in terms of the values of k_{mn} of eqn. 4.57. For values of $\epsilon_r k_0^2$ close to k_{mn}^2, the electric field (mn)th term becomes infinite whereas in practice it will be limited by the losses of the antenna. The dielectric loss may be included directly by using the complex form of ϵ_r, but some way of including the conductor and radiation losses is also required. This is then achieved by modifying the dielectric loss to represent all of the losses in the antenna by means of an effective loss tangent

$$\tan \delta_{eff} = \left(\frac{P_r + P_d + P_c}{P_d}\right) \tan \delta \tag{4.61}$$

An iterative procedure is then used where the fields are initially calculated using the true value of $\epsilon_r k_0^2$; the radiation, copper and dielectric losses are then calculated to give a modified $\epsilon_r k_0^2$ through eqn. 4.61 and the procedure repeated. The comparisons reported by Richards *et al.* (1979) then show more consistent agreement between the theoretical and experimental loci of impedance as shown in Fig. 4.19(b).

4.5.2 Corrections for fringe field leakage

The simple models described in the earlier Sections of this Chapter modelled the patch antenna as a resonator with magnetic conductor walls at the periphery, and it was shown that this model was useful for predicting many of the characteristics of the antennas. In reality, the near-field of any microstrip structure leaks out into the dielectric surrounding the conductor pattern and also out into the air region and around the substrate. This spreading of the field is known as the field fringing effect, and it must be taken into account if the accuracy of theoretical modelling is to be improved. Carver (1979) has discussed the single-mode analysis method applied to a rectangular patch with an edge wall admittance having conductance and susceptance components; however since the simple model of the antenna is so useful for radiation and circuit parameter calculations it is desirable that the allowances made should be in the form of modifications to the parameters used in the model rather than any complication of the model itself.

Some aspects of the effects of fringe fields have been discussed previously in Chapter 2 in connection with the properties of microstrip transmission lines, and it was shown that the line may conveniently be modelled as a closed line with electrically conducting top and bottom walls and magnetically conducting side walls. The physical parameters, line width and substrate permittivity, for the transmission line are changed to effective width and permittivity to allow for the fringe field effects with the relationships

$$\epsilon_e = \frac{\epsilon_r + 1}{2} + \frac{\epsilon_r - 1}{2} \left(1 + \frac{10h}{w}\right)^{-1/2} \tag{4.62a}$$

$$w_e = \frac{120\pi h}{Z_m \sqrt{\epsilon_e}} \tag{4.62b}$$

The model has been applied to a number of problems of interest such as transmission line junctions and discontinuities. Similarly, fringe field effects at discontinuities such as open-circuited transmission lines have been represented by capacitive load susceptances or effective open-circuit positions as well as taking account of radiation effects. This type of behaviour was discussed in Section 4.1.1. in relation to the conditions at the edge of a semi-infinite parallel plane system with a plane wave incident upon it within the plates.

Another viewpoint may be taken of the transmission-line characterisation of fringe field effects which is of use when the microstrip structure has an area rather than a linear nature, and the planar dimensions are of the order of a wavelength or less. In these circumstances, particularly when the patch has dimensions which are small compared to a wavelength, the analysis is carried out in terms of the true static capacitance compared to the fringe-free capacitance. This information may then be used to calculate an effective perimeter of the patch, normally equally spaced from the physical perimeter all around the patch, corresponding to an area which theoretically has the same value of static capacitance neglecting field fringing as the true capacitance of the physical structure.

Theoretical values of the static capacitance of microstrip structures have been given by a number of authors; this has been for circular discs as in Itoh and Mittra (1973) or Coen and Gladwell (1977) and for open circuits as in Farrar and Adams (1971) or Silvester and Benedek (1972). It is of interest to compare the effective position of the boundary to the microstrip line derived for the various conditions, and this has been done in Fig. 4.20. It can be seen that the two calculations based on static capacitances indicate much lower values than those obtained from the planar waveguide model when the waveguide is assumed to be filled with dielectric material characterised by ϵ_e. The planar waveguide model result may be brought into closer agreement with the other two by assuming it to be filled with dielectric material characterised by ϵ_r, as in Napoli and Hughes (1971). This assumption, however, implies that the medium has an effective permeability $\mu_e = \epsilon_e/\epsilon_r$, since the model is specified to have a propagation constant and impedance equal to that of the microstrip line, and as $\epsilon_e < \epsilon_r$ this means $\mu_e < 1$. An alternative specification for the disc model would be to require that the model should have resonant frequencies, and stored electric and stored magnetic energies equal to those of the true microstrip disc whereas the curve of Fig. 4.20 was based only upon the static stored electric field. It would be expected then that the material filling the disc model would have $\epsilon_e < \epsilon_r$, so raising the curve towards that of the planar waveguide model. An exact analysis of the disc radiator and other structures are outlined in Chapter 9 and this places the fringe corrections on a more rigorous basis.

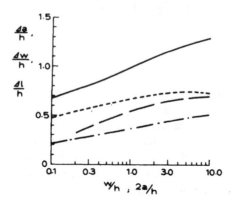

Fig. 4.20   *Comparison of effective location of microstrip magnetic wall boundaries for various configurations*
———— $\Delta w/2h$ of planar waveguide model with dielectric ϵ_e, μ_0
– – – $\Delta a/h$ of circular disc static capacitance with dielectric ϵ_r
– . – . $\Delta l/h$ of open circuit transmission line extension with dielectric ϵ_e
. . . . . $\Delta w/2h$ of planar waveguide model with dielectric ϵ_r, μ_e (< 1)
Curves shown are for substrate $\epsilon_r \sim 2.5$

It is thus apparent that although the 'magnetic wall' model of microstrip is very useful for initial engineering calculations, great care must be exercised when attempting to refine the method by making allowances for the effects of fringe fields in the form of an effective perimeter. The calculated position of such a perimeter is very dependent upon the details of the calculation procedure and may vary by a factor of more than 2:1, so there is dependence on experimental verification for a particular case. Furthermore, even if a particular procedure may be shown to give good agreement with experiment for one particular set of parameters there can be no guarantee that similarly good results will be obtained for another set of parameters. Finally, the fringing will varying with frequency as shown for the open circuit termination in Chapter 3. For practical design purposes it is useful to derive modifications to the resonant frequencies of fundamental mode disc and rectangular antennas as follows.

(i) Disc antennas: For the disc antenna, the modified resonant frequency calculation is based on an adjustment which is derived from the relationship of the true static capacitance to that of the fringe-free model. Itoh and Mittra (1973) and Wolff and Knoppik (1974) have used this type of technique based on the results of numerical analysis, but this is of limited engineering use since no analytical formula was given. Shen *et al.* (1977) have given a quasi-theoretical formula for any dielectric based on analogies with a previous result by Wolff and Knoppik (1974) for the air-filled capacitor. Thus, a first-order approximation to the static capacitance of a microstrip disc is

$$C_1 = C_0 \left(1 + \frac{2h}{\pi \epsilon_r a} \left\{ \ln\left(\frac{\pi a}{2h}\right) + 1.7726 \right\} \right) \tag{4.63}$$

where C_0 is the capacitance of the fringe-free model. The first-order resonant frequency is then given as

$$f_{r1} = f_{r0} \left(1 + \frac{2h}{\pi \epsilon_r a} \left\{ \ln\left(\frac{\pi a}{2h}\right) + 1.7726 \right\} \right)^{-1/2} \tag{4.64}$$

where f_{r0} is the resonant frequency of the simple magnetic wall model given by eqn. 4.29. Comparison with experimental results gave an estimated error of less than 2·5%, whereas the simple model gave errors of up to 17% depending on ϵ_r and h/a.

(ii) Rectangular antennas: In this case a formula may be derived starting from the network model of the rectangular antenna as discussed in Section 4.4.1. The first-order calculation assumes that the antenna can be regarded as a transmission line of dielectric constant $\epsilon_e(w_2)$, Fig. 4.21, as defined by the formula of Schneider eqn. 4.62a, and is a half-wavelength long taking into account the effective extension of the line. An expression can be derived for the total effective length of the antenna by equating it to the capacitance equivalent width of a transmission line whose width is equal to the length of the antenna; Napoli and Hughes (1971) adopt this method thus

$$l(w_1) = \frac{h}{Z_m(w_1)\beta(w_1)\epsilon_0\epsilon_r} \tag{4.65}$$

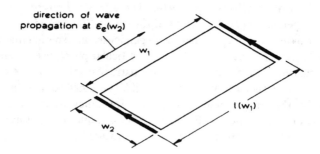

Fig. 4.21   Refinement to calculation of resonant frequency of rectangular patch using effective boundary location.

where $\beta(w_1), Z_m(w_1)$ are the phase velocity and impedance of the microstrip transmission line of width w_1. This may be rewritten

$$l(w_1) = \frac{Zh\sqrt{\epsilon_e(w_1)}}{Z_m(w_1)\epsilon_r} \qquad (4.66)$$

At resonance,

$$l(w_1) = \frac{c}{2f_{r1}} \cdot \frac{1}{\sqrt{\epsilon_e(w_2)}} \qquad (4.67)$$

Combining all these expressions and eqn (2.4) leads to

$$f_{r1} = f_{r0} \frac{\epsilon_r}{\sqrt{\epsilon_e(w_1)\epsilon_e(w_2)}} \frac{1}{(1+\Delta)} \qquad (4.68)$$

where

$$\Delta = \frac{h}{w_1}\left[0.882 + \frac{0.164(\epsilon_r-1)}{\epsilon_r^2} + \frac{(\epsilon_r+1)}{\pi\epsilon_r}\left\{0.758 + \ln\left(\frac{w_1}{h}+1.88\right)\right\}\right]$$

$$\epsilon_e(w) = \frac{\epsilon_r+1}{2} + \frac{\epsilon_r-1}{2}\left[1 + \frac{10h}{w}\right]^{-1/2}$$

and f_{r0} is the resonant frequency given by the simple magnetic wall model eqn. 4.21. Comparison with experimental results shows that this formula gives the resonant frequency to less than 3% error independent of the h/w_1 and h/w_2 ratios, whereas the simple model gives errors up to 20%.

It must again be pointed out that although eqns. 4.64 and 4.68 give first-order expressions for the resonant frequency which show the main characteristics of the dependence of the frequency upon the antenna dimension ratios Fig. 4.22(a) and (b), the residual errors will still be too large in most cases to avoid iterative development of particular antennas by experiment.

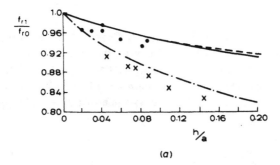

(a)

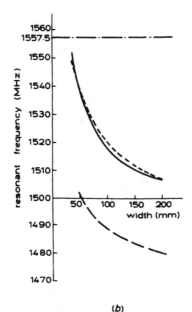

(b)

Fig. 4.22 *Comparison with experiment of theoretical resonant frequency of patches calculated using allowances for fringe fields*

a Circular patch, after Shen *et al.* (1977).
Experimental data. x $\epsilon_r = 1 \cdot 014$, ● $\epsilon_r = 2 \cdot 45 - 2 \cdot 56$
Theoretical curves: $- \cdot -$ $\epsilon_r = 1 \cdot 014$
$\qquad\qquad\quad$ —— $\epsilon_r = 2 \cdot 45$
$\qquad\qquad\quad$ $- -$ $\epsilon_r = 2 \cdot 56$
b Rectangular patch, after Lo *et al.* (1979)
Experimental curve: ———
Theoretical curves: $- - -$ eqn. 4.68
$\qquad\qquad\qquad\quad$ $\cdots\cdots$ theory given by Lo *et al.* based on experimental result
for 200 mm width
$- \cdot -$ uncorrected magnetic wall model
Patch length 59·9 mm, $\epsilon_r = 2 \cdot 62$

4.6 Short-circuit patches

The patch antennas discussed so far in this Chapter have consisted solely of the basic structure having no connections between the antenna conductor pattern and the ground-plane of the microstrip substrate. The flexibility of the antenna configuration may be increased by use of short-circuits, either of a continuous sheet form or by means of pins. Fig. 4.23 shows some of the possible variations in this

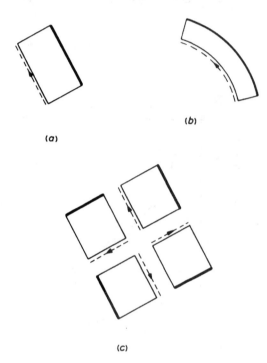

Fig. 4.23 *Microstrip patch configurations incorporating short-circuits*
a $\lambda_m/4$ rectangular patch
b $\lambda_m/4$ curved patch
c 'Crossed slot' arrangement of four $\lambda_m/4$ patches

type of antenna and particular configurations have been developed by Sanford and Klein (1978) and Garvin *et al*. (1977). The short-circuit rectangular patch antenna will be considered in more detail here since it has a particular advantage in terms of increased E-plane beamwidth compared to the open-circuit form on the same substrate material. The E-plane beamwidth may also be increased for a simple open-circuit patch by increasing the dielectric constant of the substrate, Sanford and Klein (1979), but this will lead to a decrease in bandwidth in accordance with eqns. 4.45 to 4.48.

4.6.1 Radiation patterns

This beamwidth increase obtained is due to the fact that, as noted in Section 4.2.1, the fundamental mode E-plane pattern of the open-circuit patch has essentially the interferometer characteristics of a cophased pair of sources located at each end of the patch. The half-wave open-circuit patch operating in that mode has a zero-voltage line across its centre, Fig. 4.24, and if a short circuit is added at this point

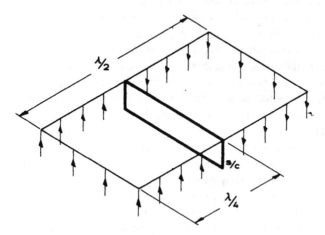

Fig. 4.24   *Relationship between $\lambda_m/2$ open-circuit and $\lambda_m/4$ short-circuit patches*

and one half of the patch removed the field structure within the resulting quarter-wavelength short-circuit patch is unchanged. The E-plane pattern then becomes that of a single magnetic current line source located at the high voltage edge, leading to an omnidirectional radiation pattern rather than one with a beamwidth of about $100°$. Eqn. 4.26 may be simply modified by eliminating the interferometer terms to give the copolarised principal plane patterns as

$$E_\theta(\phi = 90°) = -j\,\frac{\exp(-jk_0 R)2V_0 a}{2\lambda_0 R}$$

$$E_\phi(\phi = 0°) = -j\,\frac{\exp(-jk_0 R)}{2\lambda_0 R}\,2aV_0\cos\theta\,\frac{\sin\left(\dfrac{k_0 a}{2}\sin\theta\right)}{\dfrac{k_0 a}{2}\sin\theta} \qquad (4.69)$$

It must be noted that the cross-polarised radiation due to the magnetic current sources located at the sides of the patch are more significant in this case than for the open-circuit patch since each side now consists only of a half-sine distribution. Thus, whereas the $E_\phi(\phi = 90°)$ component is still zero due to cancellation between the two sides, the $E_\theta(\phi = 0°)$ has an antiphase interferometer-type pattern with a null at broadside.

4.6.2 Resonant frequency modification using shorting pins

The resonant frequency of the quarterwave short-circuit antenna using a continuous sheet as a short will be identical to that of the corresponding open-circuit antenna as discussed in previous Sections. One problem with this type of short, however, is that it normally requires a long slot to be cut in the substrate material, since the short-circuit will not usually be at the edge of the antenna board. This may require a machining operation to achieve the necessary accuracy, and at low frequencies may lead to some mechanical weakening of the substrate board. The alternative method of producing a short-circuit edge is to use a number of pins connected between the antenna conductor and ground planes, Fig. 4.25(a), and this method in addition to avoiding the difficulties noted above can easily be adapted to short circuits lying on a curved edge as described by Garvin *et al.* (1977). This will, however, produce a residual inductance at the edge rather than the desired perfect short circuit, so that for a specified resonant frequency the patch will need to be shortened slightly.

The edge effect due to the inductive component of the pins may be calculated by assuming that since the pins will produce a low electric field at that edge of the patch, there will be little fringing effect and the problem may be approximated by an array of posts across a parallel plate transmission line backed by an open circuit, Fig. 4.25(b). The equivalent circuit of an array of posts with an incident plane wave in free space has been described as a CLC T-network by Marcuvitz (1951), and this model may be applied to the arrangement of Fig. 4.25(a) by open-circuiting the output arm of the T-network as in Fig. 4.25(c). In addition, the wavelength λ' assumed in the expressions for the reactances must be that of the dielectric medium ϵ_r. The net reactance of the termination is thus $j(X_a - X_b)/Z'$ and assuming that this is a small quantity the effective position of the short-circuit is given by

$$\Delta l = \frac{\lambda'}{2\pi} \frac{(X_a - X_b)}{Z'} \tag{4.70}$$

Marcuvitz gives expressions for X_a and X_b when a/λ' is small which may be reduced for normal incidence on circular posts to

$$\frac{X_a}{Z'} = \frac{a}{\lambda'} \left[\ln\left(\frac{a}{2\pi r}\right) - 0.601 \frac{a^2}{\lambda'^2} \right] \tag{4.71a}$$

$$\frac{X_b}{Z'} = \frac{a}{\lambda'} \frac{4\pi^2 r^2}{a^2} \tag{4.71b}$$

These expressions may be substituted into eqn. 4.70 to give

$$\frac{\Delta l}{a} = \frac{1}{2\pi} \left\{ \ln\left(\frac{a}{2\pi r}\right) - \frac{4\pi^2 r^2}{a^2} + 0.601 \frac{a^2}{\lambda'^2} \right\} \tag{4.72}$$

where

$$\lambda' = \lambda_0/\sqrt{\epsilon_r}$$

The r/a and a/λ' components of this expression are plotted in Fig. 4.25(d), and it can be seen that the frequency dependent term will be negligible for all practical cases. It can be seen that the predicted extension is zero if $r/a = 0\cdot105$, which if a typical pin diameter of 2 mm were used would imply a spacing of only about 10 mm. This may, however, lead to an unacceptably high number of pins being required in some cases, and since the effective short position is quite sensitive to changes in r/a significant detuning will occur if allowance is not made for this effect.

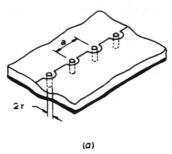

(a)

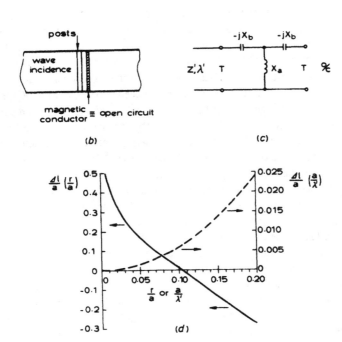

(b)

(c)

(d)

Fig. 4.25 *Network model of short-circuit composed of pins through substrate*
a Pin configuration
b Parallel-plate approximation
c Equivalent circuit
d Parameters of short-circuit as function of dimensions
$(\Delta l/a = \Delta l/a\ (r/a) + \Delta l/a\ (a/\lambda'))$.

4.6.3 Bandwidth and resonant impedance

The impedance characteristics of the quarter-wave short-circuit patch are related to those of the corresponding open-circuit patch by the following points:

(i) The stored energy within the shorted patch, for an equal peak voltage at the edge, is halved relative to the open patch since the field distribution is identical over half the area.

(ii) Only one radiating source is present in the case of the shorted patch, so the radiation conductance referred to the edge voltage is given by eqn. 4.38. Since the slot is isolated, the conductance is not modified by the mutual coupling effects which in the case of the open patch give an increase in the conductance of about 25%.

In total, this means that the input impedance of the shorted patch at resonance will be approximately 2·5 times as great as that of the open patch. This means that for the same edge voltage, the shorted patch will radiate about 40% of the power of the corresponding open patch, but as noted above the stored energy is reduced to 50% as well. Considering the basic formula for Q-factor given in eqn. 4.10 it is found that the bandwidth of the shorted patch will be reduced to 80% of that of the open patch.

4.6.4 Bandwidth increase using parasitic elements

It is possible to achieve an increase in the bandwidth of a rectangular patch system by the use of parasitic elements (Wood, 1980), two examples of suitable configurations being shown in Fig. 4.26(a) and (b). In each case, a parasitic quarter-wave shorted patch is added to the basic element so that the radiating edges of the driven element and the parasitic element are adjacent to each other. Capacitive coupling between these high voltage edges then leads to excitation of the parasitic element, and a simple circuit model is given for the first configuration in Fig. 4.26 (c). The basic circuit is equivalent to a coupled pair of resonators, which will have an even and an odd mode of resonance as shown in Fig. 4.26(d). In the even mode, the coupling capacitance and resistance do not load the circuit, and the resonant frequency of the even mode will be that of the individual circuits. In this mode the resistance, representing the radiated power of the microstrip antenna, dissipates no power and the mode is therefore not useful for an antenna. In the odd mode, however, the capacitance loads the resonant circuits so as to lower the resonant frequency by an amount which will depend upon the relative circuit element values. The radiation resistance now has a voltage across it equal to twice that of the individual circuits and will therefore radiate strongly.

The value of the radiation resistance is the same as that of the isolated edge source, since the two elements are closely spaced. Thus the power radiated by the driven/parasitic system is four times that of the isolated driven element, whereas the stored energy is only doubled with equal voltages at the edges. Thus the bandwidth of the odd mode resonance will be twice that of the isolated driven element. An identical argument may be applied to the case of the half-wave open-circuit patch with two parasitic elements of Fig. 4.26(b), and the bandwidth is again doubled. In each case, the parasitic element's radiating aperture is colocated with

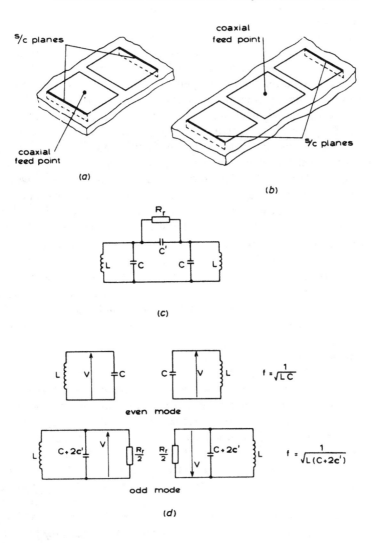

Fig. 4.26   *Parasitic short-circuit patch configurations*
   a Short-circuit driven patch
   b Open-circuit driven patch
   c Equivalent circuit of configuration shown in (*a*)
   d Even and odd resonance modes of equivalent circuit

those of the driven element, so that the radiation patterns are unchanged. The input impedance at the odd mode resonant frequency will be $\frac{1}{4}$ of the input impedance of the isolated driven element; this change is the inverse of that found with the dual-analogue folded half-wave dipole.

4.7 Summary comments

The operation of patch microstrip antennas has been described in terms of the simple 'magnetic wall' model, and it is found in practice that this is adequate for giving a general description of the properties of the antennas. However, when accurate data is required, particularly in terms of prediction of resonant frequency of the antennas, this simple model gives rise to errors and methods of improvement are required. These have been chosen so as to retain the simple basic model form, and consist of extending the analysis to include the presence of several modes, modifying the model parameters to allow for the effects of field fringing and attempting to include some modelling of the feed system. However, even with these extensions to the theory it is apparent that reliable predictions of better than one-percent of resonant frequency are difficult to achieve, and the designer will still have to resort to some experimental development to achieve a particular specification. One difficulty in adapting the simple model is that of deciding which is the best of the various methods that have been proposed, and it may be concluded that available experimental data does not give a basis for a consistent decision. Whilst the microstrip structure is not easily amenable to exact analysis, some progress is currently being made in this respect as will be discussed in Chapter 9 and the information generated also gives a better basis for assessing the suitability of the various simplified models.

One problem which has not so far been touched upon in this Chapter is that of the tolerances of substrate materials. This is a particular problem in the design of microstrip antennas which commonly use printed-circuit board materials since typical manufacturing specifications on substrate permittivity and thickness are 1·5% and 4%, respectively, although individual boards would normally be expected to be more consistent over their own areas. Variations in the permittivity are particularly important since they enter into the calculations of resonant frequency, and thus errors induced by materials tolerances are likely to be of the order of one percent which is significant compared to typical microstrip antenna bandwidths. This sort of problem must be borne in mind by a design engineer when deciding how much detail to build into the theoretical models used in design, since it may be preferable in practice to build in trimmable tuning stubs which may be adjusted individually during manufacture. Some more detailed consideration of the effects of the various types of tolerance is given in Chapter 8.

Finally, it is noted that although the main consideration in this Chapter has been given to the commonly used rectangular and circular patch antennas, many variants are possible. However, if the patch is constrained to be open-circuit, the performance obtained is very similar whatever the detailed shape as seen by comparing the results for the square and circular patches. The extension of the structure to include antennas with shorting planes or pins gives more flexibility in the control of pattern shapes etc. although the use of pins may be difficult at higher frequencies as the required dimensions will become very small. The patch antenna is therefore likely to continue to be one of the main forms of microstrip antennas in use.

4.8 References

AITKEN, J. E. (1976): 'Swept frequency microwave Q-factor measurement', *Proc. IEE,* **123**, pp. 855–862

ANGULO, C. M., and CHANG, W. S. C. (1959): 'The launching of surface waves by a parallel plate waveguide', *IRE Trans.*, AP-7, pp. 359–368

BYRON, E. V. (1972): 'A new flush mounted antenna element for phased array applications'. Proceedings of 1970 phased array symposium (Artech House, New York)

CARVER, K. R. (1979): 'A modal expansion theory for microstrip antennas'. IEEE International Symposium on Antennas and Propagation Digest, pp. 101–104

COEN, S., and GLADWELL, G. (1977): 'A Legendre approximation for the microstrip disc problem', *IEEE Trans,* MTT-25, pp. 1–6

DERNERYD, A. G. (1977): 'A network model of the rectangular microstrip antennas'. IEEE International Microwave Symposium Digest, San Diego, California, USA, (IEEE, New York), pp. 438–44

DERNERYD, A. G. (1978): 'A theoretical investigation of the rectangular microstrip antenna element', *IEEE Trans.*, **AP-26**, pp. 532–535

FARRAR, A., and ADAMS, A. T. (1971): 'Computation of lumped microstrip capacitances by matrix methods – rectangular sections and end effects', *IEEE Trans.*, MTT-19, pp. 495–497

GARVIN, C. W., MUNSON, R. E., OSTWALD, L. T., and SCHROEDER, K. G. (1977): 'Missile Base Mounted Microstrip Antennas', *IEEE Trans.*, **AP-25**, pp. 604–610

GUPTA, K. C., GARG, R., and BAHL, I. J. (1979): 'Microstrip lines and slot lines', (Artech House, New York)

HOWELL, J. Q. (1975): 'Microstrip antennas', *IEEE Trans.*, **AP-23**, pp. 90–93

IRISH, R. T. (1971): 'Elliptic resonator and its uses in microcircuit systems', *Electron. Lett.*, 7, pp. 149–150

ITOH, T. T., and MITTRA, R. (1973): 'A new method for calculating the capacitance of a circular disc for microwave integrated circuits', *IEEE Trans.*, **MTT-21**, pp. 431–432

KRETZSCHMAR, J. G. (1972): 'Theoretical results for the elliptic microstrip resonator', *IEEE Trans.*, **MTT-20**, pp. 342–343

LO, Y. T., SOLOMON, D., and RICHARDS, W. F. (1979): 'Theory and experiment on microstrip antennas', *IEEE Trans.*, **AP-27**, pp. 137–145

LONG, S. A., and SHEN. L. C. (1981): 'A theoretical experimental investigation of the circularly polarised elliptical printed circuit antenna'. Proceedings of the IEE International Conference on Antennas and Propagation, York, England, pp. 393–396

MARCUVITZ, N. (1951): 'Waveguide handbook', MIT Radiation Laboratory Series, (McGraw-Hill, New York)

MUNSON, R. E. (1974): 'Conformal microstrip antennas and microstrip phase arrays', *IEEE Trans.*, **AP-22**, pp. 74–78

NAPOLI, L., and HUGHES, J. (1971): 'Foreshortening of microstrip open circuits on alumina substrates', *IEEE Trans.*, **MTT-19**, pp. 559–561

NOBLE, N. (1958): 'Methods based on the Weiner-Hopf techniques', (Pergamon Press Ltd., London)

OLTMANN, H. G. (1978): 'Electromagnetically coupled microstrip dipole antenna elements'. Proceedings 8th European Microwave Conference, Paris, pp. 281–285

RICHARDS, W. F., LO, Y. T., and HARRISON, D. D. (1979): 'An improved theory for microstrip antennas', *Electron. Lett.*, **15**, pp. 42–44

SANFORD, G. E., and KLEIN, L. (1978): 'Recent developments in the design of conformal microstrip phased arrays'. IEE Conference on Maritime and Aeronautical Satellites for Communication and Navigation. IEE Conf. Publ. 160, London, pp. 105–108

SANFORD, G. G., and KLEIN, L. (1979): 'Increasing the beamwidth of a microstrip radiating element'. IEEE International Symposium on Antennas and Propagation Digest, pp. 126–129

SHEN, L. C., LONG, S., ALLERDING, M. R., and WALTON, M. D. (1977): 'Resonant frequency of a circular disc printed circuit antenna', *IEEE Trans.*, **AP-25**, pp. 595–596

SILVESTER, P., and BENEDEK, P. (1972): 'Equivalent capacities of microstrip open circuits', *IEEE Trans.*, **MTT-20**, pp. 511–516

WOLFF, I., and KNOPPIK, N. (1974): 'Rectangular and circular microstrip disc capacitors and resonators', *IEEE Trans.*, **MTT-22**, pp. 857–864

WOOD, C. (1980): 'Improved bandwidth of microstrip antennas using parasitic elements', *Proc. IEE*, Pt.H, **127**, pp. 231–234

Chapter 5

Linear array techniques

From initial appreciation of the possibility of direct radiation from microstrip circuits and an examination of basic radiation sources and their feeding requirements, one is led to ask whether useful linear and two-dimensional arrays can be made from microstrip.

In reviewing the response to this question by designers over the last two decades, a historical perspective of the field can be gained that allows a clearer view of the present position; possible future trends may also be seen, but discussion of these is deferred to Chapter 10. Before the 1960s, several linear array antennas had been suggested (Gutton and Bassinot, 1955; Dumanchin, 1959), but lack of suitable low loss materials prevented their development. Around this time, disc radiators were developed but the authors know of no references to arrays of resonant elements until 1972 (Byron). At this time the resonator arrays were fixed beam, and the series fed arrays of line discontinuities could be frequency scanned if necessary. During the 1970s, series-fed arrays were developed, mainly in Europe, for fixed beam communications and radar applications where the lightweight, flat profile and simple construction outweighed the reduced performance compared to conventional antennas. Arrays of resonators were also developed during this period, mainly in the USA, for similar applications although having the possibility of integrating printed phase shifters to allow beam scanning (Cipolla, 1979). Increasing interest is being taken in both forms for a wide variety of applications and the requirement for improved performance is producing a close scrutiny of both design and manufacturing problems involved. Techniques for linear arrays are examined in this Chapter and for two-dimensional arrays in Chapter 6. In the preceding Chapters, some problems particular to microstrip antennas were identified, such as bandwidth limitations, loss and surface waves; the effect on array performances of such problems is addressed here.

In this Chapter, after reviewing the forms of feeding arrangements for linear arrays, the practical forms of microstrip linear arrays that have been reported are surveyed. For this, it is convenient to split the field into arrays of resonators, arrays formed by discontinuities in a continuous line and higher-order mode arrays. In order to discuss analysis and design techniques for linear arrays, two examples are taken, namely the disc array fed through the substrate by a feed network mounted behind the ground plane, and the combline array, an example of an integral feed-array structure. These examples represent the classes of array that have the feed network, respectively, behind and coplanar with the array and represent considerably different challenges to the designer. In addition, the feed types are generally corporate and series type, respectively. In the first, the feed network will be influenced by the radiating elements in terms of element impedance and inter-element coupling. Thus a scattering parameter approach is used in the array analysis. In the second class, the feed system is further constrained by the geometry and has to fit in the space between the radiating elements. Furthermore as microstrip is an open structure, the feed system will also radiate and perturb the radiation pattern. Here the first order analysis incorporates the transmission-line matrix approach.

5.1 Review of feed methods for linear arrays

In this Section, the methods for distributing power to a linear array are reviewed. The methods are common to all linear arrays and it is necessary to survey them to grasp the significance and limitations of the feeds adopted for microstrip arrays. It should be noted that the methods discussed can be applied to both feed systems located behind the array and those formed on the same substrate. Thus in this section no distinction is made. However, comments on the space usage are made primarily from an integral feed array viewpoint; the problem of space availability may be considerably eased in rear mounted feed systems.

5.1.1 Series feeding

The simplest form of feed system for linear arrays is series feeding in which the radiating elements are attached periodically to a transmission line; the equivalent circuit is shown in Fig. 5.1. The complex loading $G_r + jB_r$ is determined by the radiating element used. The spacing between the radiating elements, L, is chosen to

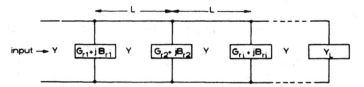

Fig. 5.1   *Equivalent circuit of series fed linear array*
L = element spacing, $G_r + jB_r$ = radiating element admittance, Y, Y_L = feed line and load admittance, respectively

to produce the required phase distribution across the aperture. In order to taper the aperture distribution, the value of G_r will change down the array, which is usually accompanied by a change in B_r which must be compensated for in the element spacing. If

$$Y_L = Y \qquad\qquad (5.1)$$

where Y_L and Y are the load and line admittance, respectively, then a travelling-wave array is formed. In this case, the exact calculation of the values of the radiation conductance G_r to produce a required amplitude distribution, given how much power can be lost in the load, P_L, requires an n-variable optimisation procedure and is very involved if losses in the line or the susceptance loadings B_r are significant. However, for long arrays, it is sufficiently accurate to perform a reverse iteration procedure starting at the load. The radiation conductance of the ith element, normalised to the line admittance Y_L, is

$$G_{ri} = P_i \left[P_L + \sum_{\nu=i}^{n} P_\nu \right]^{-1} \tag{5.2}$$

where P_i is the power radiated by the ith element and the array has n elements. The approximation involved here will result in a very small increase in sidelobe level which is usually masked by tolerance effects. P_L is, in practice, usually chosen to be 5–10% of the input power. If the values of G_r calculated are outside the range available from any particular element, then considerably more power may have to be lost in the load. This may then limit the choice of radiating element available for any application.

The phase of the radiating elements is determined by their spacing along the transmission line, therefore as the frequency is changed a progressive phase shift down the array results which causes the main beam direction to change. This is an important and limiting characteristic of the travelling wave array. To achieve a broadside directed beam the elements must all be in phase which implies one wavelength spacing. However, if this permits grating lobes, for instance with the use of an air-spaced TEM transmission line, then a reduced spacing is desirable. This results in more elements being used and allows the power to be bled from the transmission line faster or permits smaller G_r values to be used. If the polarity of the feed to the radiating element can be reversed, for example in the case of a dipole array, by switching over the feed wires or by offsetting a broadwall waveguide slot on alternate sides of the centre line, then the element spacing can be reduced to half a wavelength in the transmission line. However, in both these cases it can be seen that the reflected waves from the mismatch of consecutive radiating elements will return to the input in phase and for long arrays will produce a high input VSWR. For this reason most travelling wave arrays are operated a few degrees off broadside. If the radiating elements can be spaced by a quarter of a wavelength and alternate pairs fed in antiphase, then these reflections can be made to cancel out. This is termed quadrature feeding. This may not be possible if the elements occupy a significant physical length of the transmission line, as in the case with waveguide slots. Alternatively, small tuning stubs may be placed on the transmission line to cancel the reflections. Printed construction techniques offer much scope for these cancellation techniques and most microstrip travelling-wave arrays and also the sandwich wire antenna can operate on broadside. To the author's knowledge no broadside pointing waveguide travelling wave arrays have been made apart from circularly polarised versions which use an inherently matched pair of radiating elements to form circular polarisation.

If line losses are particularly significant, then some improvement in the efficiency can result by centre feeding the array as shown in Fig. 5.2. Increased efficiency occurs because the input power is now inserted where the amplitude distribution is maximum. However, if there is significant power lost in the load then centre feeding

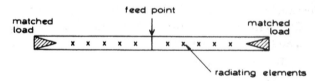

Fig. 5.2   *Schematic diagram of centre fed linear array*

may increase this as the array lengths are now halved. A particular problem with this type of feeding is that each half of the array has waves travelling in opposite directions; hence the beam from each will have a different squint angle which will severely limit the bandwidth obtainable from this arrangement. The bandwidth of the end-fed array will be limited only by the degradation of the sidelobes and the transmitted power fall off as the beam scans away from the desired direction. The input VSWR of all travelling-wave arrays is limited by tolerance factors as the periodicity will inherently cancel the element mismatches, particularly in long arrays.

If

$$Y_L = 0 \quad \text{or} \quad \infty \tag{5.3}$$

then a resonant array is formed which will have a broadside pointing beam with either wavelength or half wavelength element spacing. The input impedance will now limit the bandwidth, which is for typical resonant waveguide arrays less than 1%. Here, although no power is lost in a load, if the line loss is particularly significant, as in the case of microstrip, the saving in efficiency will be limited. Increased efficiency can be obtained, as in the travelling-wave case, by centre feeding and this is the preferred option for the resonant array. The achievable amplitude taper is only limited by the available range of G values and experience suggests that low sidelobes can be achieved more easily in a resonant array.

Beam squint with frequency is a particular disadvantage of travelling wave arrays, unless the effect is accentuated by increased line lengths for frequency shift scanning antennas. In principle, this beam movement can be eliminated by equalising the path lengths between the input and each element. This in general results in a bulky corporate feed system. However, a compact form of corporate feed has been developed, known as the series compensated feed, this was described in Section 1.1.2 and shown in Fig. 1.4.

5.1.2 Parallel feeding

The corporate or parallel feed system is simply a device that splits power between n output ports with a prescribed distribution while maintaining equal path lengths from the input to output ports. This may take the form of an n-way power splitter or a combination of m-way power splitters where $m < n$. For linear array feeding the n-way power splitter poses problems in maintaining equal phase lengths but this can be relieved by introducing some m-way splitters. A choice of $m = 2$ appears to be reasonably optimum and minimises the ratio of maximum to minimum impedance required in the feed structure. Fig. 5.3 shows two examples of corporate

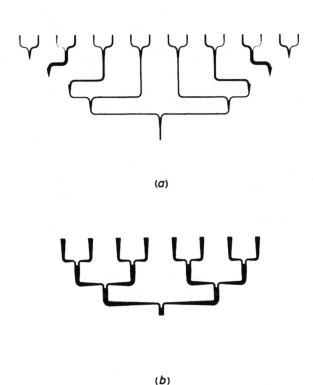

(a)

(b)

Fig. 5.3   *Corporate feed silhouettes*
 a Feed giving a distribution for -32 dB sidelobe level: $m = 2, n = 16$
 b Feed giving a uniform aperture distribution: $m = 2, n = 8$

feeds suitable for linear arrays for different amplitude distributions. The change in feed geometry between Fig. 5.3(a) and Fig 5.3(b) whilst maintaining equal path lengths highlights the flexibility of this feed structure using the two-way ($m = 2$) splitter. If broadband impedance transformers, smooth bends and splitters are used then this feed will be capable of wideband operation. The bandwidth will primarily be limited by the match of the radiating elements. This dependence on the match of the loads can be reduced by the use of isolating splitters and this is described more fully in Section 6.4.

In contrast to the series fed type, the corporately fed linear array in planar form will occupy a much larger area which will preclude its stacking to form a two-dimensional array. In practice even the series compensated feed system cannot easily be stacked.

The advantages and disadvantages of these various feed methods are summarised in Table 5.1.

Table 5.1   *Assessment of linear array feed types*

Feed type	Main beam direction	Possible bandwidth	Bandwidth limitations	Efficiency	Space usage
1. Series					
1a. Travelling wave:					
end-fed	frequency sensitive, broadside possible	wide	pattern control	moderate to high — limited by loss and available G values	efficient
centre-fed	broadside	narrow	pattern control	moderate to high — better than end fed	efficient
1b. Resonant:					
end-fed	broadside	very narrow	VSWR	moderate to high	efficient
centre-fed	broadside	very narrow	VSWR	moderate to high	efficient
2. Series compensated	broadside — frequency independent	wide	pattern control or element match	moderate to high	moderate
3. Corporate	broadside — frequency independent	wide	pattern control or element match	moderate to high	poor

It should be noted that for the series compensated and corporate feed types the beam pointing direction can be freely chosen, although for many applications broadside is preferred. In addition the efficiency of any feed type will be crucially dependent on the loss of the transmission medium used.

5.2 Practical forms of linear microstrip array

The broad classifications of linear microstrip arrays are presented here. First, linear arrays of resonant elements are considered and then the various ways in which the discontinuity array concept has been implemented are given. Finally, arrays which use radiation from higher-order modes are described.

5.2.1 Arrays of resonators

Microstrip patches can be used to form linear arrays provided that they are small enough to allow an element spacing that inhibits grating lobes. Thus rectangular halfwave-length patches (Sanford, 1978) and discs (Parks and Bailey, 1977) are used. The former are fed by a corporate feed on the same board and incorporate phase shift networks to electronically scan the beam by about $\pm40°$. The latter are fed from the rear of the board by a triplate phase shift network. Byron (1972) has joined the edges of each element in an H-plane array of rectangular patches to form a strip array, which is shown in Fig. 5.4. Each resonant cell is formed by rivets;

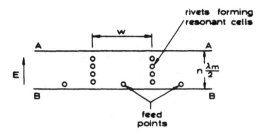

Fig. 5.4   *Microstrip strip radiators, with isolating rivets and multiple feed points*

Munson (1974) has shown these rivets are unnecessary if the strip is corporately fed, as in Fig. 5.5, with

$$N_F > L_D$$

where N_F is the number of feeds and L_D is the length of the strip in wavelengths in the dielectric. The bandwidth of such arrays will be limited by the elements used and pattern control will be governed by the precision achieved in the corporate feed design. Series-fed linear arrays of half-wavelength patches have been described by Derneryd (1975). Forming a resonant array of four such patches, Derneryd found that the bandwidth was less than for a single patch, being about 1·5% for a 2:1 VSWR for a $\lambda_0/40$ substrate height. This form of array gives a broadside beam and

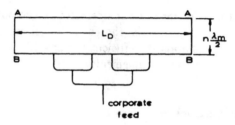

Fig. 5.5   *Microstrip strip radiator with corporate feed*

is shown in Fig. 5.6(a); if an off-broadside beam or frequency scanning is required the last resonator (Nth) can be matched to form a travelling-wave type array (Danielson and Jorgensen, 1979) shown in Fig. 5.6(b). A beam scan rate of ±30° in ±3% frequency change at 9·6 GHz was achieved at the expense of 6·6 dB loss in the

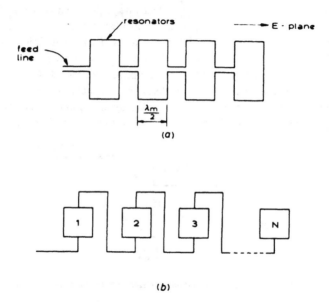

Fig. 5.6   *Directly coupled linear arrays of microstrip resonators*
   a Resonant series feed arrangement
   b Travelling-wave arrangement for swept frequency scanning

transmission lines. If very widely scanned beams are required, then half-wavelength elements are unsuitable due to the deep nulls in their radiation pattern close to the ground plane. In this case, quarter-wavelength shorted patches (Garvin *et al.*, 1977), modified disc-tab elements (McIlvenna and Kernweis, 1979) or wavelength patches can be used. However, the latter two elements may be too long for a simple straight array and some form of staggered array may be necessary.

Series feeding of resonator arrays is possible by capacitively coupling the

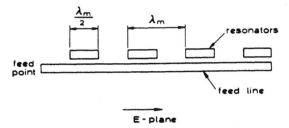

Fig. 5.7 *Capacitively coupled linear array of microstrip resonators*

resonators to the feed line (Cashen *et al.*, 1970) as shown in Fig. 5.7. The array may then be resonant or travelling wave, although no examples of the latter form are given in the literature possibly due to the high power that would be lost in the load due to the difficulty of obtaining high values of coupling to the resonators especially off the resonant frequency. The coupling to each resonator is controlled by the distance of the element from the line. In the resonant form the elements are placed at voltage maxima on the line and the element spacing in either case is one wavelength. For a practical resonant uniformly distributed array, −12 dB sidelobes are obtained but the far out sidelobe level agreed much better with theory when troughs were cut in the substrate close to the open circuits to suppress the surface wave (James and Wilson, 1977). The % bandwidth of the resonant form is input VSWR controlled and is equal to about a tenth of the beamwidth in degrees.

By rotating the resonators by 90° and directly coupling one end to the feed line, the comb line array (James and Hall, 1977) is formed. The resonators or stubs are spaced by a wavelength in the feed line so that a cophase aperture results as shown in the resonant array example in Fig. 5.8(a). The radiation conductance thrown across the feed line is directly governed by the stub width and this is shown in Fig. 5.8(c) where a tapered amplitude distribution is obtained resulting in sidelobes of about −18·5 dB, Fig. 5.8(b). Losses in this example are about 2·6 dB.

The power radiated per unit length can be increased by placing half-wavelength spaced stubs on alternate sides of the feed line which again produces a cophase aperture. Fig. 5.9(a) shows a C-band resonant array using this configuration. For such thick stubs, the T-junction data used was inaccurate but Fig. 5.9(b) and (c) shows that the dominant behaviour is predicted. Loss in the strip and the open circuit feed line were 0·4 dB each and the cross-polarisation on boresight was −25 dB. With suitable matching the bandwidth for VSWR < 1·4 was found to be 100 MHz. Experiments also showed that the bandwidth increased with substrate height to a limit which is comparable to that of air-spaced dipoles (Hersch, 1973).

The resonant array is advantageous in that no matched load is required but these examples show that this can be counterbalanced by higher values of strip loss; the bandwidth also decreases steeply with the number of stubs. The losses and bandwidth capabilities of travelling wave arrays can be assessed from the results for the array shown in Fig. 5.10(a) (Hall and James, 1978). Here a matched load is attached to one end; also in order to allow operation with a broadside beam,

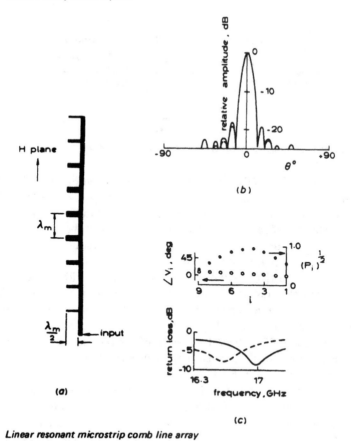

Fig. 5.8 *Linear resonant microstrip comb line array*
    a Comb line array for $\epsilon_r = 2.32$, $h = 0.793$ mm
    b H-plane radiation at 17 GHz
    ——— experimental
    – – – computed
    When $\theta = 90°$ the input is nearest the illuminating antenna
    c Return loss (relative to 50 Ω)
    ——— experimental
    – – – computed,
    ooooo phase distribution ⎫
    ●●●●● amplitude distribution ⎬ computed
    V_i and P_i (eqn. 5.2) are the *i*th element voltage and power loss, respectively. Element
    1 is nearest the input

quarter wavelength spacing is used with pairs of elements on alternate sides of the
feed line. Fig. 5.10(b) shows the return loss and sidelobe suppression obtained. The
return loss is better than 10 dB and sidelobes lower than −20 dB for a bandwidth of
about 700 MHz. The peak in return loss at the broadside frequency is due to coupling between the stubs and is discussed in Section 5.4.4. About 0·5 dB power is lost
in the load and 1·3dB in the feed lines.

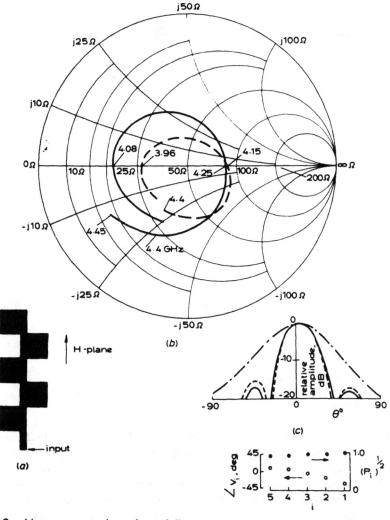

(b)

(c)

(d)

Fig. 5.9 *Linear resonant microstrip comb line array*
 a Comb line array for $\epsilon_r = 2.32$, $h = 1.59$ mm
 b Input impedance
  ——— experimental
  – – – computed
 c H-plane radiation pattern at 4.4 GHz
  ——— experimental
  – – – computed
  —·—·—· experimental E-plane pattern
 When $\theta = 90°$ the input is nearest the illuminating antenna
 d ○○○○○ phase distribution   ⎫ computed
  ●●●●● amplitude distribution ⎭
 V_i and P_i (eqn. 5.2) are the *i*th element voltage and power loss, respectively. Element
 1 is nearest the input

(a)

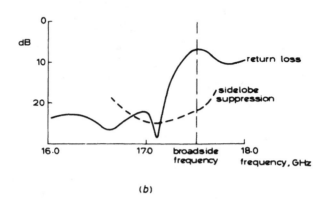

(b)

Fig. 5.10   *Linear travelling-wave microstrip comb line array*
a Comb line array for $\epsilon_r = 2\cdot32$, $h = 0\cdot793$ mm (quadrature feeding)
b Measured return loss (relative to 50 Ω) and sidelobe suppression

Comb lines have also been made at 36 and 70 GHz (Hall *et al.*, 1978) and as these took the form of two-dimensional arrays their performance is described in detail in Section 6.2.2. The conclusions given there indicate that provided the substrate height and array dimensions are suitably scaled then similar performance to that at 17 GHz can be obtained within the limits imposed by manufacturing and material tolerances, with the exception of reduced efficiency due to increase conductor losses. At 70 GHz, useful linear arrays can also be constructed using high dielectric constant substrates such as alumina because, at these frequencies the radiation conductance of the open end, which at 17 GHz is too low to allow efficient antennas to be made, is significantly increased.

The rate at which the beam direction changes with frequency can be increased in the comb line array by placing slits in the stubs as shown in Fig. 5.11 (Aitken *et al.*, 1979). The slits force the wave to travel up and down each stub thereby greatly increasing the effective line length down the array. A scan rate of $3\cdot4°$ per % frequency change is obtained which is about 3 times greater than a normal travelling wave array.

A further array formed from open circuit stubs is the 'quasi-snake line' array in which groups of nonresonant stubs of length $< 0\cdot25$ λ_m are modulated in a form

Fig. 5.11   *Swept-frequency scanned microstrip comb line array*

similar to the serpent line described in the next Section; this is shown in Fig. 5.12. This array is a derivative of the comb line but the nonresonant stubs allow a much greater range of values of power coupled from the feed line into the radiated field. A best sidelobe level of −24 dB has been achieved although broadside operation is not possible.

Fig. 5.12   *Quasi-snake line microstrip linear array*

A coupling method that simultaneously provides a wide range of coupling and a wide bandwidth is the overlaid resonator technique (Oltman, 1978), described in more detail in Section 10.2.2. Here the resonators are etched on an upper substrate having no ground plane, and mounted on a conventional microstrip substrate which has the feed network on it, as shown in Fig. 5.13. The coupling is controlled by the overlap of the resonator and the feed line. The advantages of this method are the reduction in unwanted feed radiation, if the lower substrate is thin or alternatively increased resonator bandwidth if an increased overall height is used. These improvements are offset by the added complexity and height of the assembly.

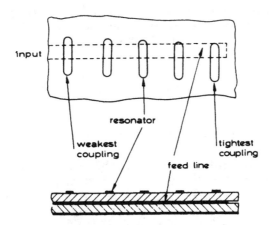

Fig. 5.13   *Linear array of overlaid microstrip resonators*

5.2.2 Array formed from line discontinuities
The fact that any discontinuity in microstrip will radiate has already led to several forms of linear array being constructed. It is, as yet, unclear whether any particular

form leads to better radiation control than another. Substrate surface waves are generated in addition to radiation and together with tolerance effects, limit the achievable sidelobe level. Achievable sidelobe levels are quoted where possible and from these it appears using currently available materials and production techniques that a highest sidelobe level of between −25 and −30 dB is achievable. Polarisation control is only provided by the rampart line or the chain antenna. In general other forms are linearly polarised.

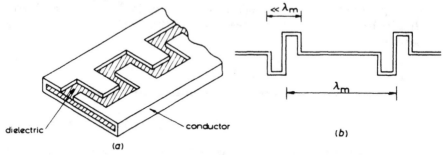

Fig. 5.14   *Microstrip line discontinuity arrays (Gutton and Baissinot)*
a Interleaved finger surface-wave structure
b Array formed by bends in microstrip line

In proposing the concept of forming linear arrays from discontinuities in microstrip lines, Gutton and Baissinot (1955) described several forms of such arrays, two of which are shown in Fig. 5.14. Impedance discontinuities in the surface wave array of Fig. 5.14(a) are formed by varying the geometry of the structure in particular finger length, width or spacing. Similarly, in Fig. 5.14(b) the discontinuity is formed by a series of bends in the line. Dumanchin (1959) has formed the discontinuity by offsetting short sections of the line as in Fig. 5.15. In a centre-fed resonant form, a sidelobe level of −25 dB was achieved. In these three forms the

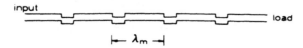

Fig. 5.15   *Microstrip line discontinuity array (Dumanchin)*

discontinuities are localised and radiation can be considered to be originating from these discrete sources. Trentini (1960) was first to propose a continuous form of discontinuity which was later called the serpent line. The microstrip line is modulated in a sinusoidal form as in Fig. 5.16(a). Wood (1979) has shown that the radiation may be modelled as an effective magnetic current M_e, which produces radiation from a curved microstrip line, and which is given by (Section 7.3.1)

$$M_e = V k_0 w_e \left(\frac{1}{k_0 R} + j \cos \phi \sin \theta \right) \tag{5.4}$$

where V is the microstrip line voltage, k_0 is the free space propagation constant and θ and ϕ are spherical co-ordinates. This indicates that radiation is inversely proportional to the radius of curvature R and proportional to the equivalent line width w_e. Thus, in the serpent line, radiation is concentrated at the peaks of the sinusoid and is controlled by the modulation amplitude. The wavelength of the modulation, arranged so that points X and Y on Fig. 5.16 radiate approximately in phase, must account for the change in path length with modulation amplitude and also susceptive effects.

Trentini has presented trapeziodal and zigzag forms as shown in Fig. 5.16(b) and (c), respectively. Measured results for lines mounted $\lambda_0/10$ above a ground plane showed sidelobes of between -20 dB and -25 dB over a narrow band of frequencies with polarisation transverse to the line.

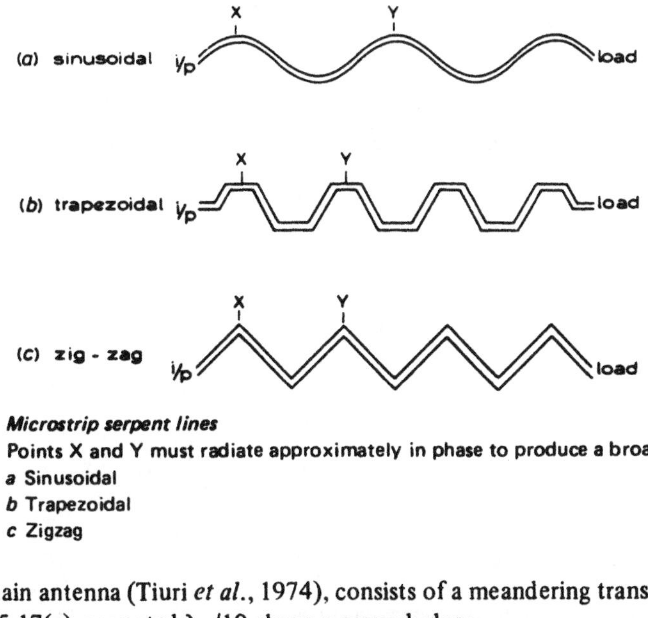

Fig. 5.16 *Microstrip serpent lines*
Points X and Y must radiate approximately in phase to produce a broadside beam
a Sinusoidal
b Trapezoidal
c Zigzag

The chain antenna (Tiuri *et al.*, 1974), consists of a meandering transmission line as in Fig. 5.17(a), mounted $\lambda_0/10$ above a ground plane.

The operation is described by Tiuri in terms of currents in the loops of these conductors. The transverse currents being in the opposite direction effectively cancel and the longitudinal currents add in the far field to produce directional radiation polarised parallel to the array length. Examples constructed at about 2 GHz, with air as dielectric with an exponentially tapered distribution produced -10 dB sidelobes with a cross-polarisation of < -30 dB. Fig. 5.17(b) shows a circularly polarised chain antenna (Henriksson *et al.*, 1979). This is similar to the zigzag serpent line but uses phase shifting elements at alternate vertices. Various combinations of α and s are possible that give circular polarisation. Coupling within the phase shifters also provides some control over the rate of beam scan with frequency although the limitations of this have not been investigated. A detailed description is given in Section 7.4.2.

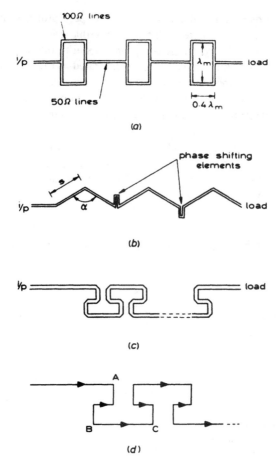

Fig. 5.17   *Microstrip chain and Franklin antennas*
    a Linearly polarised chain antenna
    b Circularly polarised chain antenna
    c Franklin antenna
    d Current distribution on Franklin antenna

A further meandering line antenna is the microstrip Franklin antenna (Nishimura *et al.*, 1979) shown in Fig. 5.17(c). Derived directly from the free space zigzag Franklin antenna (Jasik, 1961) which consists of linear array of dipoles coupled by half-wavelength lines, the action of the microstrip form is again described in terms of the currents on the line. When the line lengths AB and BC are half-wavelength, the current distribution is as in Fig. 5.17(d), resulting in a radiated field polarised to the array length. A uniform line width example has −13 dB sidelobes at 9·4 GHz with −33 dB cross-polarisation.

The rampart line (Hall, 1979; Wood *et al.*, 1978) is a meandering microstrip transmission line consisting of right-angled corners which individually produce diagonally polarised radiation but when phased correctly can form an array pro-

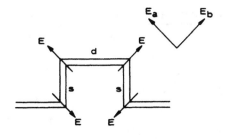

(a)

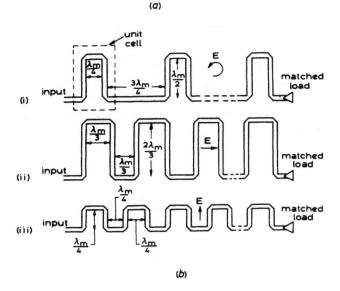

(b)

Fig. 5.18   *Linear microstrip rampart line array*
  a Unit cell showing radiated field vector at each bend
  b Rampart line forms for
      (i) circular
      (ii) parallel
      (iii) perpendicular
         polarisation

ducing variable polarisation. Fig. 5.18(a) shows a cell of the array and the polarisation of the radiated field from each corner. The design for circular polarisation is detailed in Section 7.4.1. Linearly polarised versions have also been made with the radiation polarised respectively parallel and perpendicular to the array length. The performance is summarised in Table 5.2. All arrays had uniform line width resulting in theoretical sidelobes of about −13 dB.

Silhouettes of the three arrays are shown in Fig. 5.18(b). All arrays were of the same length. The lower efficiency of the circularly polarised version is due to the

Table 5.2   *Summary of measured rampart line performance* $\epsilon_r = 2\cdot3, h = 1\cdot59mm$

Polarisation	Perpendicular	Parallel	Circular
Centre frequency (GHz)	4·0	5·0	3·7
Ellipticity (dB)	–	–	<2·0
Cross-polarisation (dB)	<−16	<−14	–
Sidelobe level (dB)	<−10	<−8	<−6
Bandwidth (%)	44	12	7
Return loss (dB)	<−13 except at broadside	–	<−10 including broadside
Efficiency (%)	60	–	27
Load loss (%)	16	–	50

reduced number of bends, allowing more power to be lost in the load. A circularly polarised line has also been produced at 17 GHz with a uniform line width giving a sidelobe level of about −11 dB over a 6% bandwidth with other parameters as in Table 5.2. Control of the line width (Hall *et al.*, 1981) and rampart geometry (Hall and James, 1981) are two possible means of tapering the aperture.

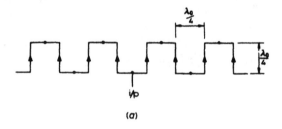

(a)

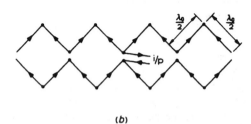

(b)

Fig. 5.19   *Wire arrays for operation in free space or* $\lambda_0/4$ *from ground plane*
Arrows show current flow when input signal has a wavelength of λ_0
a Symmetrical Bruce antenna
b Chireix-Mesny antenna

In conclusion, the physical similarity between the serpent antenna and the various forms of sandwich wire antenna, described in Section 1.3.2; suggest a close link in their modes of operation which may become obscured by the difference in approach to their understanding and analysis. In the same way the chain and rampart antennas also appear to be generically related, the linearly polarised chain being two rampart lines placed back to back. The chain antenna was derived from surface current concepts and is similar to the symetrical Bruce antenna (Kraus, 1950), which is a curtain array consisting of meandering lines in free space as shown in Fig. 5.19(a). The linear form of the free space Chiriex-Mesny array, Fig. 5.19(b), is likewise similar to the serpent line. This then suggests a common ancestry of these forms of microstrip arrays in lower frequency wire antennas, and the case of the microstrip Franklin antenna reinforces this view. The evolutionary path of the microstrip series-fed array is then as follows; a free space linear array is mounted $\lambda_0/4$ in front of a ground plane to give a unidirectional beam, the ground plane spacing is then reduced and finally a dielectric sheet introduced to support the array. Thus the linear microstrip array can be considered to be a cavity backed antenna of reduced height hence having a reduced radiation efficiency, which can, however, be increased by lengthening the antenna. This height reduction also increases the sensitivity to tolerances and the introduction of the dielectric will encourage substrate surface waves. The appropriate analysis of the structure may be done using a transmission line (James and Hall, 1977) or a surface current approach (Lewin, 1978).

5.2.3 Higher-order mode arrays

A travelling-wave array using a wide transmission line is reported by Menzel (1978),

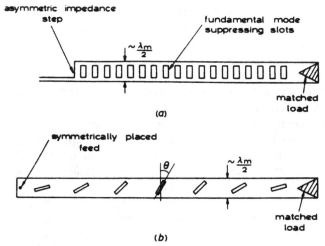

Fig. 5.20   *Higher-order mode microstrip arrays*
   a Leaky wave array
   b Slot array

and shown in Fig. 5.20(a). The transmission line is approximately $\lambda_m/2$ wide and is fed by an asymmetric impedance step which converts 80% of the incident TEM wave into the TE_{01} mode. Transverse slots are used to suppress the TEM mode in the wide line, while the remaining travelling wave radiates and forms a main beam that squints with frequency. No aperture tapering is available and the exponential aperture distribution results in sidelobes of the order of $-10\,dB$. A VSWR of $<2:1$ was obtained over a 10% bandwidth. Wood (1978) has shown that radiation occurs in a conventionally fed microstrip line loaded with slots, as shown in Fig. 5.20(b), if this line width is close to $\lambda_m/2$. The power radiated at each slot is controlled by the slot angle θ, being maximum for $\theta = 0$. Radiation from the slot and line was found to occur over a much broader bandwidth than for a conventional waveguide slot. The similarity of this array with that of Menzel suggests that the radiation mechanism may be the same and although it is unclear yet whether radiation comes predominantly from the slot, from the edges of the line, or from both, these two types form a novel and potentially useful class of microstrip antenna.

5.2.4 Overall assessment of available linear arrays

In assessing the various forms of linear microstrip arrays that are available, direct comparison is difficult due partly to the lack of information on the achievable performance limits. In this respect, it is possibly too early in the development of microstrip arrays to do this. Undoubtedly, much more work will be done here which will reveal the inherent limits of the types. Comparison is also difficult as each array has been developed from different requirements or viewpoints or simply to circumvent the patent rights taken on most of these forms. However, some general comments can be made here which may clarify the foregoing Sections:

(*i*) The performance is in general determined by the feeding methods set out in Section 5.2. In particular, corporately fed arrays have a well controlled aperture distribution and the bandwidth and polarisation are determined by the radiating elements used. Series fed arrays have a narrow bandwidth when resonant and which can be increased by operating in the travelling wave mode.

(*ii*) Arrays of resonant elements use the open circuit as the basic radiating element and, as this is the discontinuity most thoroughly analysed, such arrays are the most amenable to theoretical design.

(*iii*) Discontinuity arrays will in general be most free from unwanted feed radiation effects and may therefore be capable of better radiation pattern control than other forms. However, their design is at present largely empirical. The surface current approach may be useful in their design particularly in continuously curved structures such as the sinusoidally modulated serpent line.

(*iv*) The origins of the various forms of linear array can clearly be traced back to cavity backed forms and to slotted arrays. This suggests that microstrip as a combined transmission and radiation medium offers much flexibility in array design once the fundamental wave trapping action is understood and that many more configurations will be devised that give particular characteristics within the limitations imposed by tolerance and loss effects.

5.3 Analysis and design of rear-fed resonator arrays

Arrays formed from microstrip resonant patch elements take two forms, those that are fed from behind the ground plane, and those that have the feed network on the same substrate as the elements. In the former the feed network may be microstrip, triplate or any other form of transmission media. Here the feed performance will only be affected by the element impedance and mutual coupling and array analysis can be conveniently split into calculation of the feeding voltages and calculation of the array performance from these voltages. When the feed network is integral with the array, coupling from the array to the feed and feed radiation will occur. In some circumstances these two problems may not be significant and the dominant performance may be approximated by the isolated feed-array approach. However as the trend of array requirements is to tighter control of radiation this important problem is treated in some depth in Section 5.4 using the comb line as an example. In this Section, we will examine the disc array fed from either a triplate or micro-strip network mounted behind the array board; this example represents an important new antenna class namely narrow bandwidth, planar arrays for communication systems with the possibility of electronic scanning.

It should be pointed out that the division of arrays into hybrid and integral feed-array types is chosen only to separate out the feed radiation problem. Other problems, such as mutual coupling between patch elements, may be amenable to equivalent treatments whether the feed is isolated or integral. The reader is thus encouraged to study Sections 5.3 and 5.4 together in order to get a comprehensive view of the array design problem.

5.3.1 Element choice and first-order design

The concept of a linear microstrip array fed by a feed network mounted behind the ground plane allows any radiating element to be used that meets the required radiation specification. The feed network may then be constructed in shielded microstrip, triplate or any other suitable transmission medium and connection from the feed to the radiator is usually made by pins. This concept is illustrated by Fig. 5.21, which shows a disc array with a rear mounted triplate corporate feed.

The advantages of this form of microstrip array are primarily that the isolation of the feed from the array substrate substantially simplifies the electrical design of the array and allows much more freedom in the design of the feed network. This feed design freedom is of crucial importance in phased array applications, where the necessary digital shifters are too large to incorporate in between the elements on the microstrip substrate, and has meant that a rear fed design is the only current solution. However, the penalty paid for this freedom is increased thickness, weight and mechanical complexity.

Conventional array design methods may be applied to microstrip arrays and for a full description of these the reader is referred to, for example, Hansen (1964). The arrays considered in this and the next Chapter are planar and the principle of separability of the element pattern and the array factor applies. Thus

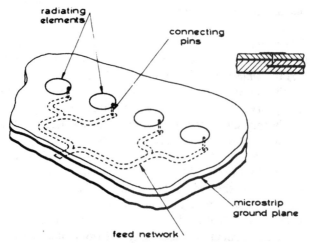

Fig. **5.21**   *Microstrip disc array with rear-mounted triplate feed*

$$E(\theta, \phi) = g(\theta, \phi) \cdot f(\theta, \phi) \tag{5.6}$$

where $E(\theta, \phi)$ is the array radiated field, $g(\theta, \phi)$ the element radiated field and $f(\theta, \phi)$ is the array factor in the direction defined by the spherical co-ordinates θ and ϕ.

The choice of radiating element and hence $g(\theta, \phi)$ for a microstrip array depends on the radiation characteristics required and the most important of these are

(i) bandwidth
(ii) polarisation
(iii) efficiency
(iv) pattern shape and scan coverage

The first two characteristics are largely independent of the array application and allow a straightforward choice of substrate dielectric constant and height, thus controlling the bandwidth, and feeding method; the latter can be used to select the correct polarisation. As most resonant elements have high efficiency, the array efficiency is largely determined by the losses in the feed system. The scan coverage determines the element spacing necessary to prevent grating lobes and if scanning close to the ground plane is required then elements with good coverage in these areas are called for.

If θ_s is the maximum scan angle required in a phased array then the maximum element spacing, d, is given by:

$$d = \frac{\lambda_0}{1 + \sin |\theta_s|} \tag{5.7}$$

The element size limits the minimum element spacing and hence θ_s and Table 5.3 shows the typical dimensions of some of the possible array elements which are shown in Fig. 5.22. It can be seen that the quarter-wavelength shorted patch is

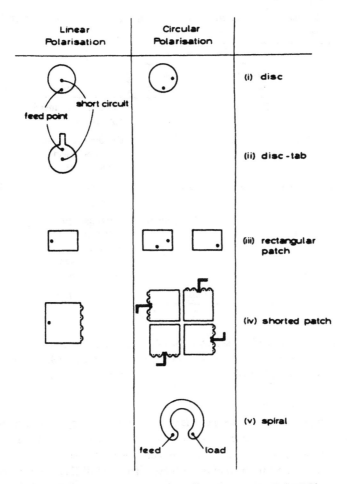

Fig. 5.22   *Types of microstrip radiating elements (not to scale: see Table 5.3)*

smallest, although the shorting elements increase manufacturing complexity. If circular polarisation is required then the half-wavelength square patch is smallest although the disc has the advantage of circularly symmetric radiation patterns, important in scanning applications. Improved element packing densities can be obtained over square arrangements shown in Fig. 5.23(a) by triangular matrix forms (Sanford and Klein, 1978) shown in Fig. 5.23(b).

Good low-angle coverage can be obtained using the quarter-wavelength shorted patch or the disc tab element. However, measurements by McIlvena and Kernweis (1979) have shown that such improvements can also be obtained by closely spacing, $(0 \cdot 4 \, \lambda_0)$, disc elements. Measurements by Bailey and Parks (1978) also show some improvements for a similar case.

It should be pointed out that improved element packing densities or the use of larger elements is facilitated by the use of substrates with higher dielectric constant

Table 5.3   *Comparison of properties of array elements for microstrip linear arrays, $\epsilon_r = 2 \cdot 32$, $h = 1 \cdot 59$ mm, frequency = 10 GHz*

| | Size | | | |
	Linear polarisation	Circular polarisation	Comment	Reference
Disc	$0 \cdot 34 \lambda_0$ dia	$0 \cdot 34 \lambda_0$ dia	Circularly symmetric patterns for c.p.	Chapter 4
Disc-tab element	$0 \cdot 51 \lambda_0 \times 0 \cdot 34 \lambda_0$		Good pattern coverage at low angles in E-plane only	Chapter 4
Half-wavelength patch	$0 \cdot 32 \lambda_0 \times 0 \cdot 32 \lambda_0$	$0 \cdot 32 \lambda_0$ square		Chapter 4
Quarter-wave-length shorted patch	$0 \cdot 18 \lambda_0 \times 0 \cdot 18 \lambda_0$	$0 \cdot 36 \lambda_0$ square	Good pattern coverage at low angles in E-plane More complex manufacture	Chapter 4
Spiral		$0 \cdot 36 \lambda_0$	Low efficiency	Chapter 7

(ϵ_r) substrates. This may also lead to reduced surface wave effects as shown in Fig. 3.16. Similarly, phase-shifter construction may be eased by higher ϵ_r substrates. This then raises the possibilities of mixed ϵ_r or high ϵ_r arrays.

5.3.2 Mutual coupling effects
Mutual coupling between radiating elements in microstrip arrays results in both distortion of the element radiation pattern and also errors in the element feeding voltages. In scanned arrays, these effects are scan angle sensitive.

Due to the open nature of microstrip these effects are difficult to calculate; this is discussed in more detail in Chapter 9. An approximate method for the disc element postulates that the electric field distribution is similar to that for the TE_{11} circular waveguide (Bailey and Parks, 1978). Thus the dual problem of coupling between dielectric filled and covered waveguide-fed apertures can be used to predict the coupling between the microstrip discs (Bailey, 1974). In this method the self- and mutual scattering parameters only are calculated; the radiation patterns are assumed to be independent of coupling and this is shown to be approximately true by measurement for discs spaced by $0 \cdot 42 \lambda_0$. Fig. 5.24 shows calculated values of the coupling between two discs for various spacings in both E-plane and H-planes.

In Fig. 5.24, coupling for the two-element array simply represents the coupling between two isolated discs for various disc centre to centre spacing. Measured values are obtained using the arrangement shown in Fig. 5.25. Removable sections of substrate allow various spacings to be obtained whilst maintaining substrate continuity. The effect on this coupling of other discs with matched outputs is shown in the results for the eight-element array. The disc spacing is fixed at $0 \cdot 42 \lambda_0$

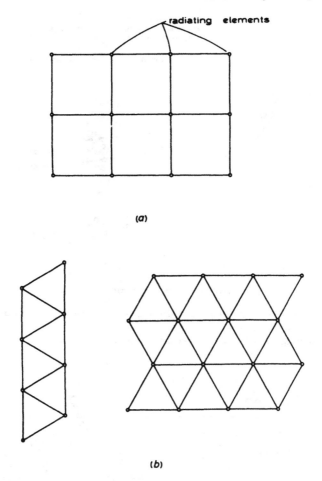

Fig. 5.23   *Element spacing arrangements for arrays of microstrip resonator elements*
    a Square
    b Triangular

and coupling is found between the first and each of the other seven elements in turn. The results show that the presence of the other discs increases the coupling. Coupling in the *E*-plane is much stronger than in the *H*-plane and Bailey and Parks believe this to be due partly to enhanced coupling to the substrate surface mode although no deductions as to the magnitude of the surface mode coupling are drawn; further comments are made in Chapter 9. Similar behaviour is predicted by analysis of dipoles on a grounded substrate, Section 9.1. Fig. 5.26 shows the effect of orientation on coupling between discs for a disc spacing of $2 \cdot 3 \; \lambda_0$.

Comprehensive measurements by Jedlicka and Carver (1979) confirm the conclusions for the disc and also show that *H*-plane coupling is stronger for the disc

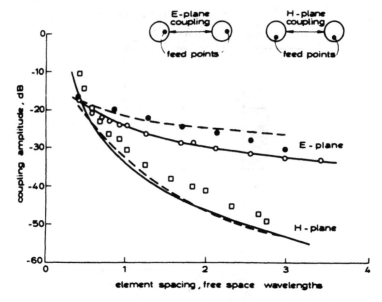

Fig. 5.24   *Mutual coupling between microstrip disc radiators as a function of spacing*
8-element array: — — — calculated; ●●● measured
2-element array: ——— calculated; ○○□□ measured

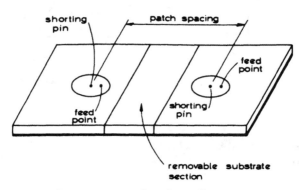

Fig. 5.25   *Arrangement for measurement of patch coupling*

than for the rectangular patch but weaker in the *E*-plane. Jedlicka and Carver have also calculated the effect of coupling on the radiation pattern of a five element uniformly fed array of patch radiators spaced by $0.5 \lambda_0$. The results are summarised in Table 5.4 and show that the change in performance is dependent on array geometry and squint angle. Krowne and Sindoris (1980) have calculated the *H*-plane coupling between rectangular patch radiators using a coupled line approach.

In calculating the radiation pattern perturbation the array is represented as an

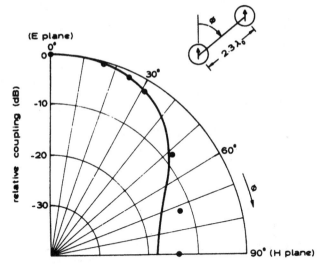

Fig. 5.26   *Mutual coupling between microstrip disc radiators as a function of orientation normalised to $\phi = 0$*
●●●●●● measured
———— calculated values based on circular waveguide fed apertures

Table 5.4   *Effect of mutual coupling on performance of linear resonator array*

	Gain loss dB	Increase in sidelobe level dB
Broadside array		
H-plane coupling	0·6	0·3
E-plane coupling	0·4	0·3
20° Squint array		
E-plane coupling	0·42	−0·15

N-port linear system, where N is the number of radiating elements, and this can be represented by a scattering matrix (Bailey and Parks, 1978). The scattering matrix represents the complex coupling coefficients between the incident (+ superscript) and reflected (− superscript) voltages at each port. Thus

$$
\begin{bmatrix} V_1^- \\ V_2^- \\ \cdot \\ \cdot \\ \cdot \\ V_N^- \end{bmatrix} = \begin{bmatrix} S_{11} & S_{12} \ldots S_{1N} \\ S_{21} & S_{22} \\ \cdot \\ \cdot \\ \cdot \\ S_{N1} & & S_{NN} \end{bmatrix} \begin{bmatrix} V_1^+ \\ V_2^+ \\ \cdot \\ \cdot \\ \cdot \\ V_N^+ \end{bmatrix} \tag{5.8}
$$

The reflection coefficient, Γ_m, of element m, is given by

$$\Gamma_m = \frac{V_m^-}{V_m^+} = \sum_{p=1}^{N} S_{mp} \frac{V_p^+}{V_m^+} \tag{5.9}$$

If the main beam is scanned in the direction θ, where θ is the angle from the normal to the array, as shown in Fig. 5.27, and $\phi = 0$ the active reflection coefficient will be

$$\Gamma_m = \sum_{p=1}^{N} S_{mp} \left| \frac{V_p^+}{V_m^+} \right| \exp[jk_0(m-p)d \sin \theta] \tag{5.10}$$

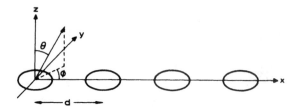

Fig. 5.27   *Geometry of microstrip disc array*

Bailey and Parks then give the excitation voltage of radiating element m as

$$V_m = V_m^+ - V_m^- \tag{5.11}$$
$$= V_m^+ [1 - \Gamma_m (\theta)] \tag{5.12}$$

which shows that the excitation voltage of each element in the array varies with scan angle due to coupling. From eqn. 5.12, the radiation pattern of the array can be found using conventional array theory.

In addition to mutual coupling, the size of the ground plane and the position of the elements on it may also be important. Bailey and Parks (1978) report that for arrays on a small ground plane this may be the dominant effect and in measurements of mutual coupling may mask the effect of direct coupling.

In Section 4.2.1, it was noted that the patch element has cross polarised radiation only in the 45° planes and at about −16 dB down on the peak copolarised level. Thus the cross-polarised radiation in the plane of the linear array will be small and the authors are not aware of any reports of this causing design problems. In two-dimensional arrays of patches cross-polarisation will be further suppressed by the multiplicative effect of the two principal plane array factors in the 45° planes.

5.4 Analysis and design of arrays with integral feed networks

The construction of microstrip linear arrays having the feed network lying on the same substrate makes full use of the space saving advantages offered by the medium. However, the combining of transmission lines and radiating elements will lead to the problem of feed-element coupling and radiation from the feed network.

The class of arrays with integral feed networks can be broadly divided into resonator arrays fed by corporate feeds and series fed arrays. Analysis and design of the former is similar to rear fed resonator arrays, Section 5.3 but the coupling and feed radiation problems in this case are difficult to handle. Indeed if these problems cause significant performance degradation then rear feeding may be the preferred configuration. First-order design methods for series arrays not involving continuously curved lines involve analysis of equivalent circuits whilst curvature may be accounted for by the equivalent magnetic current method used for the microstrip spiral, described in Section 7.3, although no analyses of curved series arrays using this method have been published.

In this Section, the analysis and design methods for series-fed linear arrays without continuous curvature are examined. First, the problem of analysis and first-order design is discussed with the linear resonator array in fixed beam and frequency scan mode and the comb line used as explanatory examples. Then theoretical and practical performance limitations are examined, using data obtained from studies of the comb line, to highlight the various problems.

5.4.1 Analysis and first-order design of series arrays

As shown in Chapter 3, the radiation characteristics of microstrip antennas may be analysed by transmission-line methods assuming that the radiation from an open circuit is confined to an aperture located at the line end or by a surface current approach where the contribution from the current over the whole strip is summed. In application to series-array analysis, the transmission-line method allows a straightforward calculation giving the dominant radiation and circuit characteristics and second-order effects such as unwanted radiation from the feed structure, mutual coupling and the effect of substrate surface waves can also be included to some extent. The surface current approach will, in principle, embrace mutual coupling and surface-wave effects, although the method has not yet been successfully applied to array configurations of interest.

In the following first-order design of series-fed resonator arrays, the transmission-line approach is used. Derneryd (1975) used the equivalent circuit, as shown in Fig. 5.28(b), of the resonator array shown in Fig. 5.28(a). Here all the resonators are of equal width, w_1, and the step discontinuity is approximated by the admittance, $G + jB$, of a microstrip open end of width, w_1. The admittances Y_1 and Y_2, corresponding to line widths w_1 and w_2, respectively, may be calculated using eqn. 2.4. Similarly, the electrical line lengths L_{e1} and L_{e2} are given by

$$L_{e1,e2} = \frac{L_{1,2}\sqrt{\epsilon_{e1,e2}}}{\lambda_0} \tag{5.13}$$

where L_1 and L_2 are the physical lengths involved, λ_0 is the free space wavelength and ϵ_{e1} and ϵ_{e2} are the effective dielectric constants given by eqns. 2.5 and 2.7. Derneryd has used this equivalent circuit to optimise the gain of such an array. Danielson and Jorgensen (1979) have used a similar equivalent circuit for each cell in a frequency scanning resonator array which uses quarter-wave matching sections

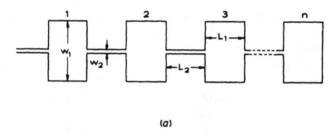

(a)

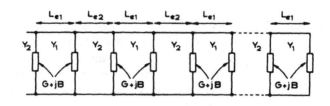

(b)

Fig. 5.28   *Series-fed microstrip resonator array*
    a Array geometry
    b Equivalent circuit

to couple in to each resonator as shown in Fig. 5.29. The discontinuity susceptances are replaced by their equivalent lengths of transmission line which means that the electrical length of the resonator is $\lambda_m/2$ at resonance. By specifying a phase-shift and attenuation constant for each cell the geometry is determined.

    The comb line is modelled using a transmission line approach (James and Hall, 1977). Fig. 5.30 shows the ith stub of the array and its equivalent circuit. In addition to radiation conductance G_r and susceptance B_r and line characterisation $A(w_i, L_i)$, $A(w, D_i)$, the ith T-junction is represented by a shunt susceptance B_{ti}, a transformer $1:n_i$ and two reference plane extensions d_{1i} and d_{2i}. The various line elements are represented by ABCD matrices and successive matrix transformation then allows the array performance to be determined.

    The computer program which embodies this analysis also determines the stub widths for any combination of sidelobe level and power lost in the matched load assuming that the substrate height, dielectric constant, feed line width and number and grouping of stubs have been previously specified. The power to be radiated by each stub, P_i, for a specified sidelobe level is calculated for a Taylor type aperture distribution (Hansen, 1964) using eqn. 5.2. The stub widths, w_i, are related to G_{ri} by

$$G_{ri} = \frac{w_i^q}{K} \tag{5.14}$$

where q and K are constants determined for a particular substrate and frequency

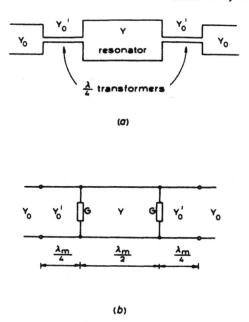

(a)

(b)

Fig. 5.29 *Swept-frequency scanning microstrip resonator array element*
  a Element geometry
  b Equivalent circuit

used, from the data of Chapter 3. In addition to analysis and design, the computer program produces a silhouette of the array on an incremental plotter for antenna production. This semi automatic method of production is typical of printed antennas and is one reason for their low production costs.

Lewin (1978) has suggested that the T-junction radiation power is about a third of the power radiated from the open circuit, the pair radiating in the manner of a half-wavelength resonator. If this is the case then this radiation cannot be considered as second order. Its effect will be to increase the loading on the feed line and this is discussed in Section 5.4.3; the E-plane beamwidth may also be affected.

5.4.2. Theoretical performance limitations of travelling-wave arrays
The performance of series-fed arrays is determined in general by the substrate used and the geometry of the radiating elements. In arrays operating in a resonant mode, the overall efficiency and bandwidth are related to that of the individual elements and there will usually be some reduction in efficiency and bandwidth involved. In a travelling-wave array the extra degree of freedom introduced by the matched load means that several performance trade-offs are possible. The results of a parametric study of the comb line (Hall and James, 1978) are shown in Fig. 5.31 and serve to highlight these trade-offs. The study was based on combs using quarter wavelength spacing with 40, 60 and 80 stubs and 0·793 mm thick substrate with $\epsilon_r = 2·32$ at 17 GHz. The bandwidth computed was always found to be sidelobe controlled and

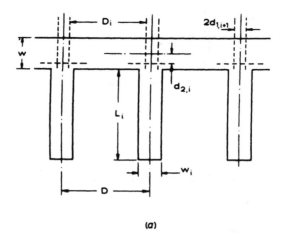

(a)

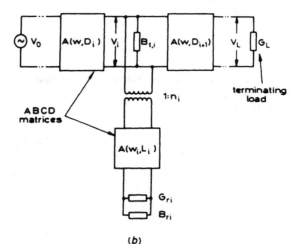

(b)

Fig. 5.30   *Microstrip comb line array*
a Array element geometry showing *i*th stub
b Equivalent circuit

very much greater than could be achieved in practice and is therefore not shown. Similarly, input VSWR was always computed to be less than 1.1:1.

The following points can be drawn from the curves:

(i) The directive gain D, Fig. 5.31(a) and (b) is related to the array length L_a, and power lost in the load P_{load} by:

$$D = K' \frac{L_a}{P_{load}} \qquad (5.15)$$

where K' is a constant. However, the absolute gain which includes line losses P_{line} will be maximum for some values of L_a depending on P_{line} and will fall off

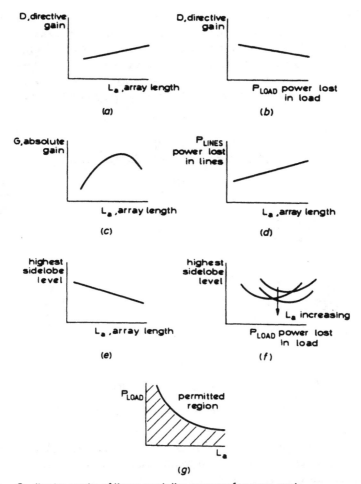

Fig. 5.31   *Qualitative results of linear comb line array performance study*

with increasing L_a as shown in Fig. 5.31(c) due to the increase of P_{line} with L_a as shown in Fig. 5.31(d). This limit on the maximum gain is discussed in Section 6.5.

(ii) If the element radiation conductance is limited by geometry considerations, as in the comb line with quadrature feeding, then the highest sidelobe level will decrease with L_a as deeper amplitude tapers are applied as shown in Fig. 5.31(e). If manufacturing and design errors are significant this may not occur in practice. The sidelobe level is minimum for some values of P_{load} as shown in Fig. 5.31(f).

(iii) There is only a certain permitted region of combinations of P_{load} and L_a, as shown in Fig. 5.31(g) due again to the finite element radiation conductance. Working close to the edge of this region, with consequent high values of element conductance may degrade input VSWR and sidelobe level due to high mutual coupling.

Limited element radiation conductance is a problem in all types of travelling wave array, although in certain waveguide slot configurations high couplings of waveguide to radiated power close to $-3\,\mathrm{dB}$ are possible. Due to the high wave trapping, it is believed that for discontinuity type arrays the maximum coupling will be significantly less than this while for resonator element types high coupling may be achieved. As an example of this limited coupling, in the comb line array having quarter-wave spacing the maximum coupling is $-16\,\mathrm{dB}$ and this is due to the limited space between the fingers. This problem may possibly be overcome by

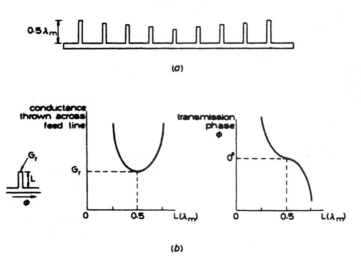

Fig. 5.32   *Stub length tapering technique for linear microstrip comb line array*
a Array geometry
b Feed line loading and transmission phase against stub length

modulation of the stub length as well as the width. Fig. 5.32(a) shows a wavelength-spaced comb line of equal stub widths with length modulation. The admittance thrown across the feed line Y is given by:

$$Y = Y_s \left[\frac{G_r + jY_s \tan \beta L}{Y_s + jG_r \tan \beta L} \right] \tag{5.16}$$

where Y_s is the stub line admittance, βL is the electrical stub length which includes the open-circuit end effect Δl and T-junction effects and G_r is the open-circuit end radiation conductance. When $\beta L = \pi/2$ then $Y = Y_s^2/G_r$, which will result in heavy line loading as $G_r \ll Y_s$. Also when $\pi/2 < \beta L < \pi$, Y will be complex and in addition to increased loading, transmission phase errors will be introduced into the feed line as shown in Fig. 5.32(b). If these errors are corrected by adjusting the stub spacing then an array with a wide range of feed line to radiation coupling can be formed.

In addition to maximising the excitation at the centre of the array it is important to have good control over the excitation at the array ends. Discontinuity arrays offer good control here as the discontinuity may be reduced smoothly to zero. However, in the combline and other directly coupled resonator arrays where coupling is proportional to line width, etching tolerances may limit the thinnest line available hence limiting sidelobe level.

A further limitation to the performance is the inherent loss in the microstrip line which is covered in detail in Chapter 2. It is worth noting here that this is an important fundamental limitation to array performance and if not severely limiting the possible applications then it may call for a reassessment of the overall system in which low cost and profile can be traded with efficiency.

The results of the parametric performance study are in general true for all series fed linear arrays although in some types much greater control of the radiation conductance is possible leading to a wider range of possible design combinations. Good control of the minimum power that can be radiated is provided by discontinuity type arrays and in this case pattern control will be limited by extraneous radiation and coupling effects. The loss problem is common to all microstrip linear arrays.

5.4.3 Design parameter characterisation effects

The parameters used in the equivalent circuit analysis of series-fed arrays for example line, open-circuit end, impedance step, T-junction and right-angle bend parameters are not precisely known and in this Section their assessment and accuracy are discussed. How each of these parameters affects array performance depends on the geometry; in general errors in conductances will affect the attenuation constant and errors in susceptive components and reference plane extensions will influence the phase constant. This Section is based on results taken from work on the comb line array but these results also serve to highlight parameter characterisation problems that are common to all series-fed arrays.

(a) Line characterisation effects: The characterisation of the microstrip line is covered in detail in Chapter 2. The basic expressions for Z_m and ϵ_e are given by Wheeler (1965) and are in two parts, for narrow and wide lines. Improvements to the crossover point between these two regions are given by Owens (1976) which lead to the maximum error and graphical discontinuity being reduced to less than 1%. These improvements were made for microstrip on alumina ($\epsilon_r \approx 10$) and when similar methods are applied to low dielectric constant substrates (Hall and James, 1978) a graphical discontinuity of about 0·7% can be achieved. Fig. 5.33 shows the errors, $\Delta Z_m/Z_m$, and, $\Delta \epsilon_e/\epsilon_e$, in the equations for Z_m and ϵ_e where $\Delta Z_m = Z_m - Z_m'$ and $\Delta \epsilon_e = \epsilon_e - \epsilon_e'$. Z_m' and ϵ_e' are accurate results obtained using Bryant-Weiss Green's functions (Cisco, 1972). The plots of ΔZ_m were calculated for $w/h > C_z$ using eqn. 2.4b and for $w/h < C_z$ using eqn. 2.4a and the more accurate formula (Owens (1976))

$$Z_m = \frac{Z}{\pi\sqrt{2(\epsilon_r + 1)}} \left[H - \frac{1}{2}\left(\frac{\epsilon_r - 1}{\epsilon_r + 1}\right)\left(\ln\frac{\pi}{2} + \frac{1}{\epsilon_r}\ln\frac{4}{\pi}\right)\right]$$

$$\text{for} \quad \frac{w}{h} < C_z \tag{5.17}$$

where C_z is the value of w/h at which the change in equation is made. The equations used for ϵ_e are (Owens (1976))

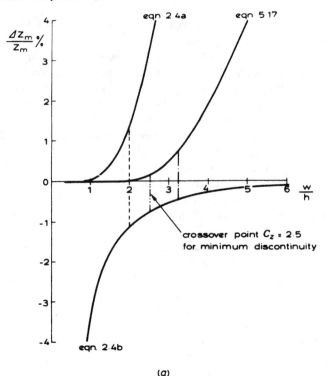

(a)

Fig. 5.33 *Accuracy of microstrip analysis expressions for $\epsilon_r = 2.32$, $h = 0.793\,mm$*
a Error curves for microstrip impedance, Z_m

$$\epsilon_e \;=\; \frac{1}{2}\left[\,\epsilon_r + 1 + (\epsilon_r - 1)\left(1 + 10\,\frac{h}{w}\right)^{-0.555}\right] \quad \text{for} \quad \frac{w}{h} > C_\epsilon \qquad (5.18)$$

$$\epsilon_e \;=\; \frac{\epsilon_r + 1}{2}\left[1 - \frac{1}{2H}\left(\frac{\epsilon_r - 1}{\epsilon_r + 1}\right)\left(\ln\frac{\pi}{2} + \frac{1}{\epsilon_r}\ln\frac{4}{\pi}\right)\right]^{-2} \quad \text{for} \quad \frac{w}{h} \leqslant C_\epsilon \quad (5.19)$$

where C_ϵ indicates the crossover value, and

$$H \;=\; \ln\left[\frac{4h}{w} + \sqrt{16\left(\frac{h}{w}\right)^2 + 2}\;\right]$$

It can be seen that for minimum graphical discontinuity $C_z = 2.5$ and $C_\epsilon = 1.0$.

Errors in ϵ_e of the order of $\pm 0.1\%$ will lead to errors in the resonant frequency of resonant arrays of $\pm 0.4\%$ and beam pointing errors in travelling-wave arrays of about $\pm 0.2°$. Such errors may be significant in very narrowband arrays or those requiring high beam pointing accuracy. Experience indicates that they are usually masked by manufacturing and material tolerance errors or errors in characterisation of other parameters. The discontinuity in ϵ_e and Z_m may also give rise to a small rise in sidelobe level in arrays that use line width tapering due to the discontinuity

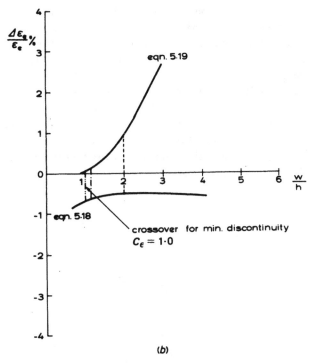

Fig. 5.33 *Accuracy of microstrip analysis expressions for $\epsilon_r = 2 \cdot 32$, $h = 0 \cdot 793mm$ (continued)*
b Error curves for effective dielectric constant, ϵ_e
— — — Wheelers (1965)
—.—.— Owens (1976)   crossover points
—..—.. Hall and James (1978)

in the phase of the aperture distribution. This is harder to quantify and again may be masked by other effects.

(b) Radiation admittance determination: Calculation of the open-circuit end parameters, $G_r + jB$ is discussed in Section 3.2, together with two ways of measurement, namely the resonator and the VSWR method. Both of these methods establish the parameters of a single, isolated open end. Due to the presence of the modifying effects of the comb line environment a method of measuring G_r in situ can provide useful data. Two possible methods exist. In the first, the transmission loss of combs having uniform width stubs is measured and G_r calculated. In the second method, the same arrays can be short-circuited and the difference in the peak levels of the main and 'ghost' beam found, which also gives a measure of G_r. The angular difference between the beams also allows the phase constant down the array to be found; this method is given in detail by Laursen (1973). The second method is only suitable for low values of G_r; for large values, the 'ghost' beam may become obscured in the sidelobes of the main beam. For this reason the first method is preferred and provided that good control over the coaxial to microstrip transition is maintained will give accurate results. Once the transmission loss of a

selection of comb lines of various stub widths have been measured, the value of G_r used in the analysis computer program is varied to produce a value of power lost in the load equal to the measured transmission loss. Measured values for $h = 0.793$ mm and 1.58 mm, $\epsilon_r = 2.32$ substrate are shown in Fig. 5.34. The spread of values for $h = 1.58$ mm substrate reflect the difficulties in control of the coax to microstrip transitions. Also shown are values for an isolated open end measured using the VSWR technique whose good agreement with comb results may suggest that environmental effects lead to only small changes in G_r. The comb line method is, however, open to errors introduced by uncertainties in the parameters used in the computer model, particularly in the correction to the stub length and spacing due to the T-junction and open-circuit end.

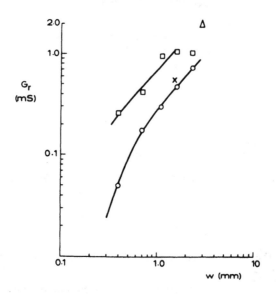

Fig. 5.34   *Microstrip open-end conductance G_r against line width w, for $\epsilon_r = 2.32$, frequency = 17 GHz*
o, h = 0.793 mm; □, h = 1.58 mm measured using comb line transmission loss method
x, h = 0.793 mm; △, h = 1.58 mm measured using VSWR method on isolated open end

Lewin (1978) has suggested that radiation off the T-junction in the comb line is comparable to that of the open end itself and hence that line loading will be significantly increased. This suggests that in any calculation these should be included and although the above measured results indicate that this increased loading may be small, insufficient data is as yet available to draw a firm conclusion.

(c) T-junction effects: The T-junction is shown in Fig. 5.35(a) together with its equivalent circuit as shown in Fig. 2.9(e). In Appendix B, expressions for the

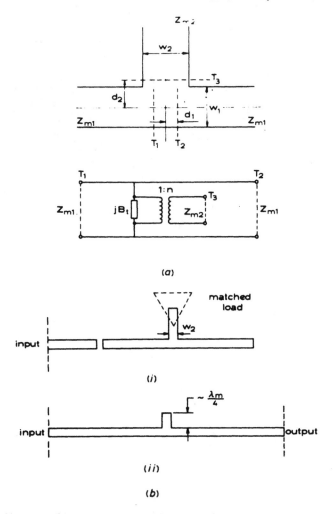

Fig. 5.35 *Microstrip T-junction parameter characterisation*
    (*a*) Junction geometry and equivalent circuit
    (*b*) Resonators used in characterisation
        (i) Line resonator for $2d_1'$, eqn. 5.22 measurement
        (ii) Stub resonator for d_2 measurement

various parameters are given (Hammerstad and Bekkadal, 1975). Hammerstad (1975) gives improved expressions for n and d_2':

$$
n^2 = \left[\frac{\sin\left(\dfrac{\pi}{2}\dfrac{2D_1}{\lambda_m}\dfrac{Z_{m1}}{Z_{m2}}\right)}{\dfrac{\pi}{2}\dfrac{2D_1}{\lambda_m}\dfrac{Z_{m1}}{Z_{m2}}} \right]^2 \left[1 - \left(\pi\,\frac{2D_1}{\lambda_m}\frac{d_2'}{D_1} \right)^2 \right] \tag{5.20}
$$

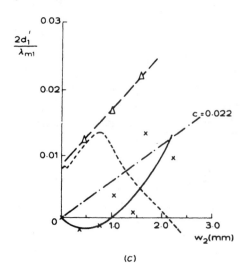

(c)

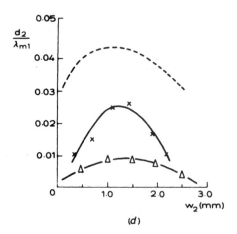

(d)

Fig. 5.35 *Microstrip T-junction parameter characterisation (continued)*
(c) $2d_1'/\lambda_{m1}$ against w_2 for $w_1 = 2\cdot5$ mm, $h = 0\cdot793$ mm, $\epsilon_r = 2\cdot32$, frequency = 17 GHz

—x— measurements on resonator i
— — — eqn. B.12, Appendix B
— Δ — modal analysis
— · — · comb design approximation eqn. 5.23

(d) d_2/λ_{m1} against w_2 for parameters as (c)
—x— measurements on resonator ii
— — — eqn. 5.21
— Δ — modal analysis

$$\frac{d'_2}{D_1} = \left[0.076 + 0.2 \left(\frac{2D_1}{\lambda_m} \right)^2 + 0.663 \exp\left(-1.71 \frac{Z_{m1}}{Z_{m2}} \right) \right.$$

$$\left. - 0.172 \ln\left(\frac{Z_{m1}}{Z_{m2}} \right) \right] \frac{Z_{m1}}{Z_{m2}} \quad \text{where} \quad d_2 = \frac{D_1}{2} - d'_2 \qquad (5.21)$$

These equations, together with eqn. B.12, Appendix B, have been used by Hall and James (1978) in the analysis of the comb line. They suggest that the single most important parameter in the equivalent circuit is d_1 as this determines the phase of the radiating elements and errors in it are cumulative down the array. For example, for a 40-stub comb having quarter-wavelength stub spacing the effect of ignoring d_1 is to produce a quadrature phase error of about $100°$, in an aperture distribution having a taper for $-30\,dB$ sidelobe suppression. This will lead to an increase in side-lobe level of the order of 5 dB. Errors in d_2 will lead to stub length errors and hence errors in the line loading combined with further smaller feed line phase errors. Errors in B_t will also give rise to aperture phase errors. Experience on the comb line indicated that the expressions for these parameters were not accurate enough for precise antenna design and therefore experimental methods were used. The resonators shown in Fig. 5.35(b) were used to measure the reference plane extensions. The stub on resonator (i) was terminated by a matched load and the resonant frequency noted for various stub widths, w_2. As the stub is placed at a voltage maximum, the shunt susceptance B_t will be excited and the change in resonant frequency will yield $2d'_1$ where

$$2d'_1 = 2d_1 - \frac{\tan^{-1}(B_t Z_{m1})}{\beta_1} \qquad (5.22)$$

where β_1 is the phase constant of the main line. The measurement result is plotted in Fig. 5.35(c), normalised to the main line wavelength, λ_{m1}. Also plotted are values calculated from eqn. B.12, Appendix B and those given by the mode matching technique (Wolff *et al.*, 1972). It can be seen that agreement is poor. In view of these apparent discrepancies together with the possible further uncertainties introduced by mutual coupling effects in the array, a combination of these results and empirical fine tuning was used for the comb line design. Here $2d'_1$ was given by

$$2d'_1 = \frac{cw_2}{(1 - cw_2)} D_s \qquad (5.23)$$

where c is a constant and D_s is the stub spacing in mm. For $D_s = 0·25\,\lambda_m$ and $c = 0·022$, $2d'_1$ is shown in Fig. 5.35(c). The optimum value of c depended on the comb geometry and was found to be in the range $0·02 < c < 0·03$.

Transmission-loss measurements on resonator (ii) Fig. 5.35(b) yielded the stub physical length for an electrical length of $\lambda_m/4$. Thus addition of a further $\lambda_m/4$ gave stub lengths for use in comb line design. Using values of λ_m and Δl found from equations 5.17, 5.18 and B.1 Appendix B, the corresponding values of d_2 again normalised to λ_{m1} can be found and these are shown in Fig. 5.35(d) compared to Hammerstad's results and those from the modal method. Again it can be seen that there are significant differences.

In general the results derived from the static and modal methods do not agree well with measurements. The difference is noted in the next chapter where these two analytic techniques, used for discontinuities in triplate, are also compared. However, in the triplate case radiation is absent and the modal method gives improved results at high frequency. In microstrip at low frequencies modal and static methods both give results in good agreement with measurement (Kompa, 1978). At high frequencies, the static approximation is less good and therefore gives poor results. The modal method, while modelling a closed waveguide discontinuity well, also becomes less useful at high frequency due to radiation. Therefore, for discontinuity characterisation with the precision necessary for the design of microstrip array antennas, where radiation is considerable, an analysis which included radiation effects is necessary.

In conclusion then, this Section has shown that parameter characterisation is a problem with all forms of microstrip linear array although the particular effects will vary with array geometry. In general, arrays using discrete radiation sources such as the open circuit and sharp bend may well be simpler to characterise than curved forms. Also, discontinuity arrays due to their integral feed line may have less problems with spurious radiation. However, there still remains much to be done in the field of characterisation and due to the lack of knowledge design of linear arrays remains to a large extent an empirical exercise.

5.4.4 Mutual coupling, unwanted radiation and cross-polarisation effects

The use of an equivalent circuit representation of series arrays means that in a first-order design, mutual coupling and extraneous radiation are initially ignored and then added later if found to be significant. How important they are will depend on the array geometry and its application. In this Section the effect on the quarter-wavelength spaced comb line is described as indicative of the magnitude and complexity of the problems involved.

Fig. 5.36 shows a typical section of comb line and the radiation and coupling which can occur between the various elements present. Using a transmission line analysis approach these effects must be added in a piecemeal way. Use of the surface current approach described earlier may allow all of these effects to be modelled simultaneously.

The close spacings involved in the double stub comb line may suggest that the value of a transmission line analysis is limited. However, as seen in Section 5.2.1, the method allows good prediction of the dominant radiation and circuit characteristics of the comb line and the application to second-order effects such as coupling, although not giving precise design details, does allow considerable insight into the action. Such second-order effect analysis starts by assuming radiation to be localised, in this case to the open-circuit end and T-junction, and then coupled circuit methods are applied. The radiating sources are represented by magnetic dipoles. If there is coupling between a pair of such dipoles having excitation

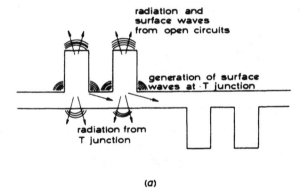

(a)

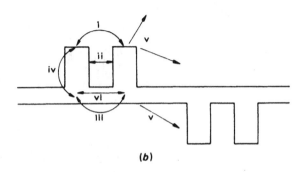

(b)

Fig. 5.36 *Unwanted radiation and coupling in the microstrip comb line array*
 a Radiation and surface-wave generation
 b Coupling within the comb line
   (i) Radiation and surface-wave coupling at open end
   (ii) Fringing field coupling
   (iii) Radiation and surface-wave coupling at T-junction
   (iv) T-junction to open-end coupling
   (v) Coupling to other parts of the structure
   (vi) Reverse coupling by dominant mode and higher-order mode coupling inside
     structure

voltages V_1 and V_2 and self- and mutual-admittances Y_1, Y_2 and Y_m, respectively, than the actual admittance of each source Y_{1A} and Y_{2A} will be:

$$Y_{1A} = Y_1 + \frac{V_2}{V_1} \cdot Y_m$$

$$Y_{2A} = Y_m \cdot \frac{V_1}{V_2} + Y_2 \qquad (5.24)$$

In general, Y_1, Y_2 and Y_m will be complex and dependent on the stub widths. If the sources represent the open ends of two adjacent stubs or adjacent T-junctions on a quarterwave spaced comb line, then

$$V_2 = jV_1$$

and the effect of coupling (i), Fig. 5.36 will be

$$Y_{1A} = G_r + jB_{r1} + jG_m - B_m$$

$$Y_{2A} = G_r + jB_{r2} - jG_m + B_m \qquad (5.25)$$

where G_r and G_m are radiation and mutual conductance, respectively, B_{r1}, B_{r2} and B_m are self- and mutual-susceptances, respectively, and it is assumed that the mutual coupling does not significantly affect the excitation voltages. The degree of

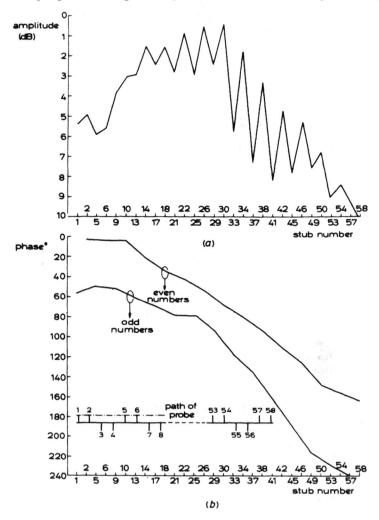

Fig. 5.37 *Amplitude and phase measured close to stub ends of microstrip comb line at 17 GHz*

approximation involved here is difficult to assess but it is believed that the dominant effect is given by eqn. 5.25. This is supported by the results of probe measurements close to the ends of the stubs, Fig. 5.37, which show some imbalance in the sampled amplitude, as suggested by the real parts of Y_{1A} and Y_{1B} in eqn. 5.25. Similarly, the different quadrature phase errors for odd and even numbered stubs suggest that a stub width dependent phase error is present that is different for the stubs in each pair as in the imaginary part of eqn. 5.25. In addition to loading and phase errors, this effect is thought to be responsible for some degradation of the comb performance on broadside. Without coupling, the quarter-wavelength stub spacing ensures that the reflection from each pair of stubs is zero and broadside operation is thus possible. The phase errors due to coupling perturb this and result in an input return loss of around 5 dB and a sidelobe level increase of between 4 and 8 dB on broadside. These internal reflections may be reduced by the inclusion of small matching stubs as shown in Fig. 5.38 and when properly tuned result in an input return loss of 12 dB and sidelobe increase of 1 or 2 dB when the beam is scanned through broadside.

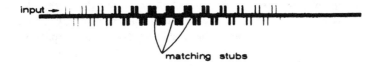

Fig. 5.38 *Microstrip comb line array with matching stubs*

Due to the symmetry of eqns. 5.25 the total radiation conductance will be as for no mutual coupling and this suggests that the effect on the feed line loading will be small. Mutual coupling between the open end and the T-junction, and T-junction radiation thus appear to represent the major perturbation to the loading. Lewin (1978) has suggested that the T-junction radiates about one third of the power radiated by the open end and the mutual coupling between these sources, assuming them to be small magnetic dipoles, is given by

$$\frac{G_m}{G_r} = \frac{3}{2\alpha} \left[\left(1 - \frac{1}{\alpha^2}\right) \sin \alpha + \left(\frac{1}{\alpha}\right) \cos \alpha \right] \tag{5.26}$$

where $\alpha = k_0 d$ and d is the source spacing. In this case

$$\frac{V_2}{V_1} = 1$$

and it is the mutual conductance, which for an $\epsilon_r = 2 \cdot 25$ substrate is $0 \cdot 31\, G_r$, that affects the line loading in eqn. 5.25. Thus the total line loading is increased, due to T-junction radiation and coupling of nearest neighbour T-junction to an open-end, by a factor of $(1 + 0 \cdot 31) \times 2 = 2 \cdot 62$.

The more general expression, for an axial separation y and equatorial separation z, is also given by Lewin:

$$\frac{G_m}{G_r} = \frac{3}{\gamma^3}\left[\left\{1 + \alpha^2 \frac{\left(1 - \frac{3}{\gamma^2}\right)}{2}\right\}\sin\gamma - \left\{1 - \frac{3\alpha^2}{2\gamma^2}\right\}\gamma\cos\gamma\right] \quad (5.27)$$

where $\alpha = k_0 z$ and $\gamma = k_0(z^2 + y^2)^{1/2}$. We have seen that the mutual coupling between sources in phase quadrature will have a small effect on the loading; the closest in-phase coupling will be between the first and third T-junction of Fig. 5.36b. Here $\alpha = 0$ and $\gamma = 0.7\pi$, giving $G_m/G_r = 0.59$. For coupling between the first and third open-circuit end, $G_m/G_r = -0.23$ and more distant coupling will be further reduced.

These results suggest that mutual coupling and spurious radiation effects will indeed be significant in the quarter-wavelength spaced comb line and may be approximately included by the above methods. In addition to line loading and phasing problems mentioned above, unwanted radiation due to surface wave reradiation and feed transitions will give rise to radiation pattern control problems. This will primarily lead to increases in the sidelobe level of the copolarised radiation; it will also however lead to significant cross-polarised lobes (Brown, 1980). The cross-polarised component radiated by the T-junction and stub will not have the phase reversal associated with the copolarised radiation from stubs on opposite sides of the feed line and will hence form a beam displaced from the copolarised main beam. Brown has measured these beams at about $70°$ to $80°$ from the normal to the substrate, at a level of -15 dB for 5 λ_m long comb line with cophased feeding, with a stub and feed line width of 8 and 5 mm, respectively. This level rose for thicker stubs. It is expected that the level of these cross-polarised beams will also depend on array length, and will be suppressed for longer arrays, as for example Hall and James (1978) have measured a level of -23 dB for a 15 λ_m long array.

5.5 Summary comments

In this Chapter, the question of whether useful arrays can be made in microstrip has been addressed, and from the plurality of existing forms the answer clearly is that useful devices can be made within the limitations imposed by the heavy wave-trapping action and the high line losses. Application is then governed by the actual values of these various limits which dictate the ultimate performance that can be achieved. Although it is evident that improved performance will be obtained by further work, the survey section and the section detailing analysis and design techniques allow the currently achieved performance and possible limits to be identified .

For rear-fed resonator arrays, bandwidth and efficiency are largely determined by individual patch performance and feed losses. These parameters may be traded by choice of resonant or non-resonant radiating elements. Sidelobe control limits are not clear yet although more than 20 dB suppression has been obtained in experimental arrays. The cross-polarisation level is determined by the element performance and may form a limit to the possible sidelobe suppression, particularly in short series-fed arrays. The presence of the dielectric sheet and possible dielectric cover may lead to blindness effects noted in phased arrays, although in the arrays made so far, edge effects and design tolerances have meant that this problem has not been encountered.

In series-fed arrays, the sidelobe level and bandwidth, and to some extent VSWR, are limited by material and manufacturing tolerances and circuit parameter characterisation accuracy. It is doubtful whether both material tolerances and design accuracy will be substantially improved in the near future and thus it does not seem probable that sidelobe levels significantly better than -25 to $-30\,dB$ will be achieved using existing methods. For similar reasons the loss problem does not seem open to much improvement using current configurations. However, in view of the low efficiency relative to other antennas forms and the present importance attached to losses this is a topic worthy of further research. The cross-polar level is limited by the polarisation purity of the radiating elements and again a judicious choice of element may help; material inhomogeneities and conductor roughness may also contribute and as little work has been done here, it is difficult to assign values to the various contributions and a possible lower limit.

The review of feed methods indicated that the general performance is largely dictated by the choice of feed and that trade offs between bandwidth, losses and space needed can be made. This together with the survey of available types indicate the wide variety of forms that can be created using microstrip. This fact clearly mirrors the large number of wire, cavity-backed and waveguide arrays available and indicates that although many of these forms have now been translated into microstrip there is still much scope for further innovation.

5.6 References

† See footnote at end of References

AITKEN, J. E., HALL, P. S., and JAMES, J. R. (1979): 'Swept frequency scanned microstrip antenna'. European Patent Application 79300 898.8

BAILEY, M. C. (1974): 'Analysis of finite size phased arrays of circular waveguide elements'. NASA Tech. Report R-408

BAILEY, M. C., and PARKS, F. G. (1978): 'Design of microstrip disk antenna arrays', NASA Technical Memorandum 78631

BROWN, A. K. (1980): 'Cross polarisation characteristics of linear comb line microstrip antennas', *Electron Lett.*, **16**, pp. 743–744

BYRON, E. V. (1972): 'A new flush mounted antenna element for phased array applications', in 'Phased array antennas'. Proc. 1970 Phased Array Symp., (Artech House, New York) pp. 187–192

CASHEN, E. R., FROST, R., and YOUNG, D. E. (1979): 'Improvements relating to aerial arrangements'. British Provisional Patent (EMI Ltd.) Specification 1294024

† CIPOLLA, F. W. (1979): 'A 7·5 GHz microstrip phased array for aircraft to satellite communication'. Proc. Workshop on printed circuit antenna technology, New Mexico State University, USA, pp. 19–1 to 19–18

CISCO, T. C. (1972): 'Design of microstrip components by computer'. NASA Contractor Report, CR-1982, pp. 7–23

DANIELSON, M., and JORGENSEN, R. (1979): 'Frequency scanning microstrip antenna', *IEEE., Trans.*, AP-27, pp. 146–150

DERNERYD, A. G. (1975): 'Linear microstrip array antennas'. Chalmers University of Technology, Division of Network Theory, Gotenburg, Sweden, TR75057

DUMANCHIN, R. (1959): 'Microstrip aerials'. French Patent Application 855234

ELLIOTT, R. S. (1959): 'Mechanical and electrical tolerances for two-dimensional scanning arrays', *IRE Trans.*, **AP-6**, pp. 114–120

GARVIN, C. W., MUNSON, R. E., OSTWALD, L. T., and SCHROEDER, K. G. (1977): 'Missile base mounted microstrip antennas', *IEEE Trans.*, **AP-25**, pp. 604–610

GUTTON, K., and BASSINOT, G. (1955): 'Flat aerial for ultra high frequencies'. French Patent 703113

HALL, P. S., GARRETT, C., and JAMES, J. R. (1978): 'Feasibility of designing millimetre microstrip planar antenna arrays'. Proc. AGARD Conf. 245 on Millimetre and Submillimetre Wave Propagation and Circuits, Munich, pp. 31–1 to 31–9

HALL, P. S., and JAMES, J. R. (1978): 'Microstrip array antennas'. Final Report on UK MOD Research Agreement AT/2160/033RL, Royal Military College of Science

HALL, P. S. (1979): 'Rampart microstrip line antennas'. European Patent Application 79301340

HALL, P. S., WOOD, C., and JAMES, J. R., (1981): 'Recent examples of conformal microstrip antenna arrays for aerospace applications'. IEE 2nd Int. Conf. on Ant. and Prop., York, pp. 397–401

HALL, P. S., and JAMES, J. R., (1981): 'Conformal microstrip antenna'. Final Report on Phase II of Research Agreement D/DRLS/5/33/11, Royal Military College of Science

HAMMERSTAD, E. O. (1975): 'Equations for microstrip circuit design'. Proc. 5th European Microwave Conference, Hamburg, pp. 268–272

HAMMERSTAD, E. A., and BEKKADAL, F. (1975): 'Microstrip handbook'. ELAB Report STF 44 A74169, The University of Trondheim, Norwegian Institute of Technology

HANSEN, R. C. (1964): 'Microwave scanning antennas'. (Academic Press, New York & London) vol. 1

HENRIKSSON, J., MARKUS, K., and TIURI, M. (1979): 'A circularly polarised travelling wave chain antenna'. Proc. 9th European Microwave Conference, Brighton

HERSCH, W. (1973): 'Very slim high gain printed circuit microwave antenna for airborne blind landing aid'. Proc. AGARD Conf. 139 on Antennas for Avionics, Munich

JAMES, J. R., and HALL, P. S. (1977): 'Microstrip antennas and arrays Pt. 2 – new design technique', *IEE J.*, MOA., **1**, pp. 175–181

JAMES, J. R., and WILSON, G. J. (1977): 'Microstrip antennas and arrays, Pt. 1 – fundamental actions and limitations', *IEE J.*, MOA, **1**, pp. 165–174

JASIK, H. (1961): 'Antenna engineering handbook', (McGraw Hill, New York) pp. 4–36

†JEDLICKA, R. P., and CARVER, K. R. (1979): 'Mutual coupling between microstrip antennas'. Proc. Workshop on printed circuit antenna technology, New Mexico State University, USA, pp. 4–1 to 4–19

KOMPA, G. (1978): 'Design of stepped microstrip components, *Rad. & Electron. Eng.*, **48**, pp. 53–63

KOMPA, G., and MEHRAN, R. (1975): 'A planar waveguide model for calculating microstrip components', *Electron. Lett.*, **11**, pp. 459–460

KRAUS, J. D. (1950): 'Antennas', (McGraw Hill, London)

KROWNE, C. M., and SINDORIS, A. R. (1980): 'H-plane coupling between rectangular microstrip antennas', *Electron. Lett.*, **16**, pp. 211–213

LAURSEN, F. (1973): 'Design of periodically modulated triplate antennas'. Proc. AGARD Conf. 139 on Antennas for Avionics, Munich

LEWIN, L. (1978): 'Spurious radiation from microstrip', *Proc. IEE*, **125**, pp. 633–642

McILVENNA, J., and KERNWEIS, N. (1979): 'Modified circular microstrip antenna elements', *Electron. Lett.*, **15**, pp. 207–208

MENZEL, W. (1978): 'A new travelling wave antenna in microstrip'. Proc. 8th European microwave Conference, Paris, pp. 302–306

MUNSON, R. E. (1974): 'Conformal microstrip antennas and microstrip phased arrays', *IEE Trans.*, **AP-22**. pp. 74–78

NISHIMURA, S., NAKANO, K., and MAKIMOTO, T. (1979): 'Franklin-type microstrip line antenna'. International Symposium Digest, 'Antennas and Propagation', Vol. 1, Seattle, Washington, pp. 134–137

OLTMAN, G. H. (1978): 'Electromagnetically coupled microstrip dipole antenna elements'. Proc. 8th European Microwave Conference, Paris, pp. 281–285

OWENS, R. P. (1976): 'Accurate analytical determination of quasi-static microstrip line parameters', *Rad. & Electron Eng.*, 7, pp. 360–364

PARKS, F. G., and BAILEY, M. C. (1977): 'A low sidelobe microstrip array'. IEE AP-S Symposium, Sanford, USA, pp. 77–80

PETRIE, E. M and GROVE, R. (1971): 'Dimensional stability of stripline materials'. Proc. 10th Elect. Insul. Conf., Chicago, pp. 179–183 (IEEE, New York)

SANFORD, G. G. (1978): 'Conformal microstrip phased array for aircraft tests with ATS-6', *IEEE Trans.*, **AP-26**, pp. 642–646

SANFORD, G. G., and KLEIN, L. (1978): 'Recent developments in the design of conformal microstrip phased arrays'. IEE Conf. on Maritime and Aeronautical Satellites for Communications & Navigation. IEE Conf. Publ. 160, pp. 105–108

TRENTINI, Von G. (1960): 'Flachantenna mit Periodisch Gebogenem Leiter', *Frequenz*, **14**, pp. 230–243

TIURI, M., TALLQVIST, S., and URPO, S. (1974): 'The chain antenna'. IEEE AP-S International Symposium, Atlanta, USA, pp. 274–277

TIURI, M., HENRIKSSON, J., and TALLQVIST, S. (1976): 'Printed circuit radio link antenna'. Proc. 6th European Microwave Conference Rome, pp. 280–282

WHEELER, H. A. (1965): 'Transmission line properties of parallel strips separated by dielectric sheet', *IEEE Trans.*, **MTT-13**, pp. 172–185

WOLFF, I., KOMPA, G., and MEHRAN, R. (1972): 'Calculation method for microstrip discontinuities and T-junctions', *Electron. Lett.*, 8, pp. 177–179

WOOD, C. (1978): Unpublished notes

WOOD, C. (1979): 'Curved microstrip lines as compact wideband circularly polarised antennas', *IEE J.*, MOA, 3, pp. 5–13

WOOD, C., HALL, P. S., and JAMES, J. R. (1978): 'Design of wideband circularly polarised microstrip antennas and arrays'. IEE Conference on Antennas & Propagation, London. IEE Conf. Publ. 169, Pt. 1, pp. 312–316

† These and many other contributions from the New Mexico Workshop have subsequently been published in a special issue of *IEEE Trans.*, **AP-29**, Jan. 1981

Chapter 6

Techniques and design limitations in two-dimensional arrays

The arraying of several linear arrays to form a two-dimensional microstrip array is a natural progression that leads to a wide variety of pencil-beam planar antennas and these are finding increasing use in many communication and radar systems today. In this Chapter, the possible feeding methods for two dimensional arrays are briefly outlined, thus forming a logical extension to Section 5.1; practical examples that have been made are then surveyed and finally analysis and design methods are examined in some depth. This brings out some limiting factors, such as coupling, surface waves and feed radiation. For large arrays with integral feeds the latter problem may necessitate the use of triplate feed structures and the design problems involved in these and microstrip feeds are addressed, including radiation from the feed to microstrip transition.

6.1 Review of feed methods

When forming a two-dimensional microstrip array from several linear arrays laid side by side on the same substrate, any of the feed methods for linear arrays described in Section 5.1 and summarised in Table 5.1 can be used in either plane. A large number of combination of feeding arrangements is possible and some of the more useful are noted in Table 6.1.

(a) *Travelling-wave feed and travelling-wave linear arrays:* This combination will produce a beam squinted in both planes. Use of the four available input ports, as in Fig. 6.1(a), will result in four squinted beams as commonly used in Doppler navigation systems.

(b) *Series-compensated or corporate feed with travelling-wave linear arrays, (Fig. 6.1(b):* This will produce a wide bandwidth array with a beam that squints with frequency in the plane of the linear arrays. Bandwidth will usually be limited by the linear array radiation pattern degradation.

(c) *Resonant feed with resonant linear arrays, Fig. 6.1(c):* These are inherently narrow bandwidth, broadside pointing arrays. As indicated in Table 5.1, centre-feeding will produce the highest efficiency as line losses will be minimised particularly for arrays with deep amplitude tapering to reduce sidelobes.

Table 6.1   *Useful combinations of feed methods for two-dimensional arrays*

Plane of linear array ╲ Plane normal to linear array	Travelling wave (end-fed)	Resonant (end-or centre-fed)	Corporate or series-compensated
Travelling wave (end-fed)	(a) Beam squinted in both places		(b) Broad bandwidth
Resonant (end-or centre-fed)		(c) Narrow bandwidth, compact feed	
Corporate (rear-mounted)			(d) Broad bandwidth, rear mounted feed

(*d*) *Corporate feed in both planes:* Room is generally not available for this type to be mounted on the antenna substrate and feeds in microstrip or triplate are formed behind the array. The network is inherently wide bandwidth unless narrow band splitters or phase shifters are used. The array bandwidth then is limited by the radiating element used.

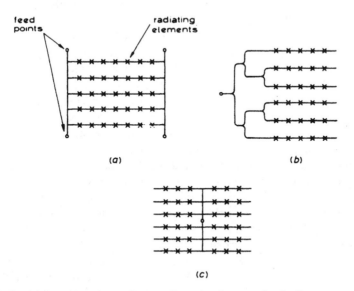

Fig. 6.1   *Feed system arrangement for two- dimensional arrays of series linear arrays*
    a Four-beam travelling-wave array
    b Corporately fed travelling-wave array
    c Centre-fed series resonant array

6.2 Practical forms of two-dimensional arrays

Two-dimensional arrays can be formed from any of the linear array elements described in the last Chapter. There are many elements available and the possibilities and limitations of many of them as array radiators are not yet fully understood. However, some resonant elements such as the disc and rectangular radiator and some forms of series arrays are now being examined in depth and it is possible to see that the fundamental issues of tolerance control, mutual coupling and unwanted feed radiation are aspects that will limit the ultimate performance obtainable. On the other hand, however, the many forms available indicate a great flexibility in the concept. The arrays that have been reported are surveyed in this section and examined in two categories, those fed by corporate feeds in both planes and those made up of arrays of series fed radiators.

6.2.1 Corporately fed arrays

Two-dimensional arrays, fed corporately, can be split, as suggested in the last Chapter, into those fed from behind and those having the feed network on the same substrate. However, the only array of the latter type, (Murphy, 1979), described more fully in Section 6.5 and consisting of 16 × 64 elements, is broken down into 64 subsections fed by a coaxial corporate feed mounted behind the substrate to reduce feeder losses.

Byron (1972) has described the development of the microstrip strip radiator. A single disc radiator, Fig. 6.2(a), can be fed $180°$ out of phase at two points. Arrays of such radiators can be formed from strips by antiphase feeding at appropriate points, Fig. 6.2(b). By separating each radiating cell by short-circuiting pins, similar operation is obtained with single feed points, Fig. 6.2(c). Using an array simulator in which one or two periods of the array are surrounded by metallic walls to simulate an infinite array environment, best match for the two-probe strip radiator was obtained with the probes at the strip edge. Measured element patterns were found to be well defined and Byron draws the conclusion that such elements may well be suitable for phased array antennas. Munson (1974) has also described a two-dimensional array formed from strip radiators. Here the strips are fed by 8-way corporate feeds in the same plane as the strips and the four corporate feeds are fed from behind the substrate as shown in Fig. 6.3. Designed for X-band operation the line losses are 0·5 dB and the uniform distribution produces sidelobes of −11 dB.

A rear-fed array of disc radiators has been made by Bailey and Parks (1978) in an array feasibility study. A section of the 8 × 8 element array, having $0·42 \lambda_0$ element spacing in a square matrix, is shown in Fig. 6.4(a). Using triplate Butler feed networks to obtain low sidelobes, sidelobe levels of less than −20 dB were achieved over an 11% bandwidth, and bandwidth degradation outside this range is attributed to feed distribution errors. The array is constructed for operation at 5 GHz on a 1·6 mm thick teflon fibre-glass substrate. The limited impedance bandwidth of the radiating elements determines the achievable bandwidth, this being 7.5%, 8% and 2% for a single disc element, an 8-element linear array and the 8 × 8

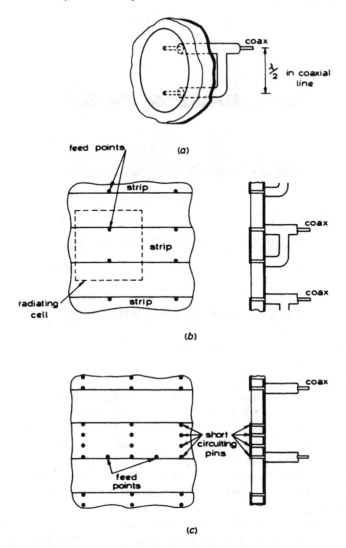

Fig. 6.2   *Microstrip strip radiator development*
   a Two-probe disc element
   b Two-probe strip radiator
   c Single-probe strip radiator

planar array, respectively. The efficiencies, including feeder losses, for the same antennas are 79%, 78% and 69%, respectively.

Beam-steering has been incorporated into an 8 × 8 element disc array (Cipolla, 1979) by feeding each element through a 3-bit phase shifter. The corporate feed, phase shifter and 90° hybrid coupler, necessary to produce circular polarisation from each disc, are produced in microstrip on a 0·5 mm thick teflon fibreglass substrate. This feed circuit, a section of which is shown in Fig. 6.4b, is bonded to the

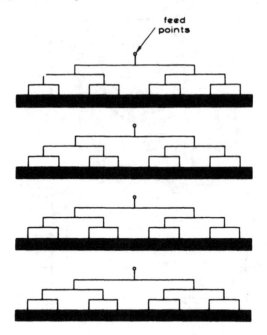

Fig. 6.3   *Two-dimensional microstrip strip array*

array substrate and connections made with pins through the substrates, as in Fig. 6.4(a). The array uses a disc spacing of $0.56 \lambda_0$ in the diagonal planes and has a square formation; the array is produced on a 0.79 mm thick substrate. Scanning is possible over a $\pm 45°$ range and the -15 dB sidelobes at broadside, produced by the corporate feed designed to give a uniform distribution, degrade to about -5 dB when the beam is scanned to $45°$. The phase shifters contribute 1.2 dB to the over-overall losses of 3.4 dB, resulting in an efficiency of about 46%. The array operates at 7.5 GHz and is aimed at a high gain, full hemisphere coverage, aircraft to satellite communication application.

A 2×16 element disc array, for mounting on a remotely piloted vehicle, is described by Yee and Furlong (1979) for operation at 0.97 GHz. Here, scanning is only required in the 16-element plane. A 32-output triplate corporate feed is mounted directly behind the array. The microstrip 4-bit phase shifters are then mounted on the rear of the triplate and connections made by through plated holes. An exploded view of a section of the array is shown in Fig. 6.5. Scanning out to close to $90°$ has been achieved, although a high grating lobe appears when the main beam is scanned out to $\pm 90°$ due to the $0.5 \lambda_0$ element spacing. Sidelobes are maintained at around -12 dB expected for the uniform distribution, out to a scan angle of $60°$. The loss in the phase shifters and feed lines is about 3.5 dB.

6.2.2 Arrays of series-fed radiators

The strip radiator of Byron (1972) can be formed into a series-fed, two-dimensional array as shown in Fig. 6.6 (Markopoulos and Catechi, 1977). The connecting feed

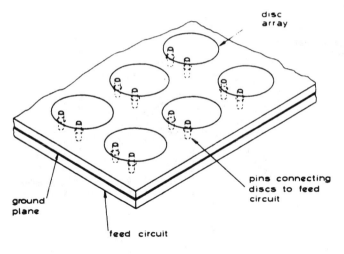

(a)

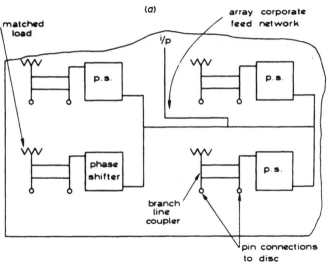

(b)

Fig. 6.4   *Microstrip phase scanned disc array*
   a Disc array arrangement
   b Feed circuit showing position of phase shift networks (p.s.)

lines are of length *t*, given by

$$t = \frac{\lambda_0}{2\sqrt{\epsilon_e}} - h \tag{6.1}$$

where *h* is the substrate height. *h*, in eqn 6.1, approximately compensates for the reactance at the transition from the feed line to the strip. In addition, short-

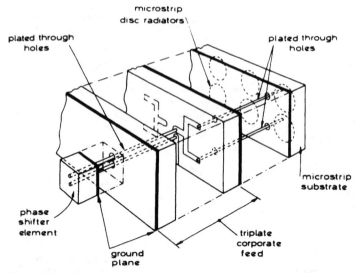

Fig. 6.5   *Exploded view of a section of a microstrip phase scanned 2 × 16 disc array*
(Substrate thicknesses are exaggerated for clarity)

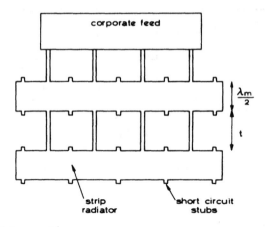

Fig. 6.6   *Series-fed array of microstrip strip radiators*

circuited stubs are used to cancel out the fringing capacitance at the strip edge. The authors indicate that the use of these stubs increases the bandwidth of a single strip radiator by a half to 15%, for $h/\lambda_0 = 0.04$ with $\epsilon_r = 4.1$. However, when a two strip array with short circuited stubs was formed, the bandwidth was reduced to 10% again.

A two-dimensional array of square resonators has been described by Derneryd (1976) for operation at 9 GHz as shown in Fig. 6.7 The 4 × 4 element planar array is formed by feeding four linear arrays in parallel. The construction allows two

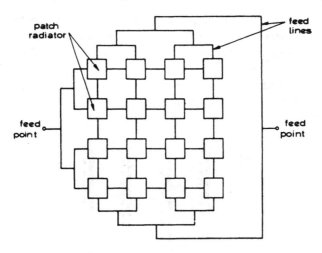

Fig. 6.7 *Series-fed array of rectangular microstrip patches*

orthogonal feeding networks to be used, to enable either horizontal or vertical polarisation to be obtained. Sidelobe levels are −10 dB and isolation between the two ports is 20 dB. The aperture distribution could be varied by altering the impedance levels of the feed lines as demonstrated by Metzler (1979). However, by a unique cross-feeding arrangement, shown in Fig. 6.8, Williams (1977, 1978) has produced an array of resonators having a tapered distribution with equal-width feed lines. Sidelobes of the order of −20 dB are achieved with efficiencies of greater than 80% for bandwidths of the order of 1 to 2%. Cross-polarisation is of the order of −20 dB.

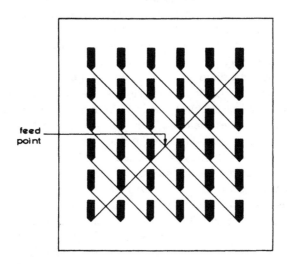

Fig. 6.8 *Microstrip cross-fed array*

Arrays of capacitively coupled resonators have been produced by Emi Varian (1973) for frequencies from 4 to 36 GHz based on the linear arrays of Cashen *et al.* (1970). A typical example is shown in Fig. 6.9. The % bandwidths are approximately (± minimum beamwidth in degrees ÷ 10), with sidelobes of −12 dB maximum. The antennas have been integrated with Gunn oscillators, mixer and detector diodes and directional couplers to form compact, low-cost transmitter and receiver radar front ends.

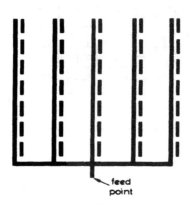

feed
point

Fig. 6.9   *Array of capacitively coupled microstrip resonant radiators*

Two-dimensional comb line arrays have been described by James and Hall (1977). A resonant example is shown in Fig. 6.10(a) designed for operation close to 17 GHz. Designed for a maximum sidelobe level of −18 dB in both planes by tapering the finger width and the linear array impedance levels, the array had a sidelobe level of −16 dB over a very narrow bandwidth of about 0.5% and the input return loss is similarly narrowband. Line losses are about 4·4 dB. Wider bandwidth performance was obtained from the travelling-wave array, Hall and James (1978), shown in Fig. 6.10(b) and (c) with −20 dB sidelobes being obtained over 7% in the *H*-plane which has an aperture amplitude taper designed for −25 dB sidelobes. In the *E*-plane, a similar distribution was used and −21 dB sidelobes were obtained over the same bandwidth. The input return loss was about 14 dB over the 7% bandwidth which was for a beam squinted up to 5° from normal to the substrate. For a normal beam, the sidelobes increased; this problem is described more fully in Section 5.4.4. Radiation from the microstrip corporate feed was found to be a significant limitation to both array gain and sidelobe level control in this array, and necessitated the use of a triplate corporate feed. Feed design and components characterisation is discussed in Section 6.4.1 and the transition from triplate to microstrip in Section 6.4.2. Losses in the feed and comb lines are 1·5 dB each which together with losses of 1·0 dB in the load and transitions produce an overall efficiency of 22% and a power gain of 26 dB, with a 5° × 5° beamwidth. Cross-polarisation is about −23 dB. The best sidelobe levels of −25 dB appear to be close to the optimum that can be achieved from this type of antenna at this frequency, in

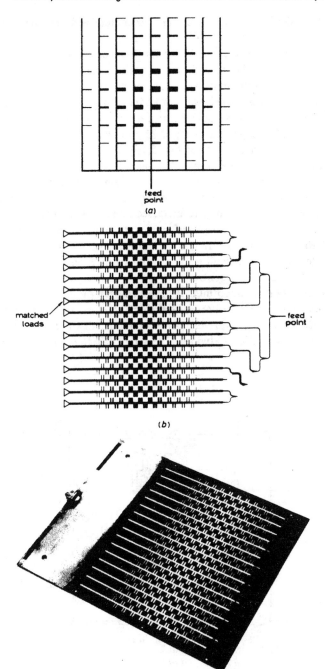

Fig. 6.10   *Microstrip comb line arrays for $\epsilon_r = 2 \cdot 32$, $h = 0 \cdot 793$ mm substrate*
a J-band resonant array
b J-band travelling-wave array with triplate feed
c Photograph of (b)

view of the tolerances both in design parameters and the manufacturing and material production processes.

Travelling-wave comb line arrays have also been designed for operation at 36 and 70 GHz (Hall *et al.*, 1978). At 36 GHz, using the same $h = 0.793$ mm substrate as was used at 17 GHz, both uniform and tapered distribution arrays were made. Sidelobes of -15 dB were obtained over a 3% bandwidth with losses similar to those at 17 GHz. However, at 70 GHz poor control of the radiation pattern indicated that thinner substrates were needed and a 0.25 mm thick substrate, the closest to an exact scaling (based on the data of Fig. 3.16) that was available, was used. For a uniform distribution, sidelobe levels of about -12 dB were obtained. An efficiency of 23% was measured. It was concluded for low dielectric constant substrates that substrate thickness scaling is necessary and that at higher frequencies some reduction in gain may occur due to increased line losses. Radiation pattern control is then limited by material and manufacturing tolerances. For alumina substrates it was found that only at 70 GHz was radiation sufficient to allow construction of efficient antennas. The difficulty of making triplate feeds in alumina, however, limited the radiation pattern control.

Series arrays of line discontinuities have also been presented in two-dimensional form. For example, the linearly polarised chain antenna with air dielectric (Tiuri *et al.*, 1976), shown in Fig. 6.11, had an exponential distribution and sidelobes of about -12 dB over a 20% bandwidth. The input VSWR was found to have a sharp

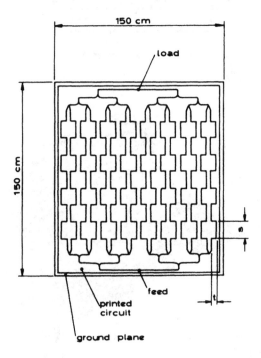

Fig. 6.11 *Linearly polarised chain antennas*

peak for a broadside beam frequency. The efficiency is not quoted, but can be deduced from the array area and gain to be about 40%, for a 5° x 5° beam. The cross-polarisation level is about $-30\,$dB. The circularly polarised chain antenna (Henriksson *et al.*, 1979) has also been used in two-dimensional form. The linear array is described in Section 5.2.2 and the two dimensional form in Section 7.4.2. It is sufficient to mention here that sidelobes of about $-11\,$dB were obtained with an efficiency of 50 to 60% for a 5° x 15° beamwidth.

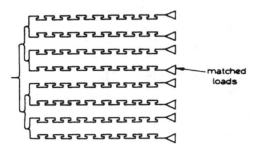

Fig. 6.12   *Microstrip Franklin array*

The microstrip Franklin antenna has been formed into a planar array (Nishimura *et al.*, 1979) as shown in Fig. 6.12. A microstrip corporate feed presents a uniform amplitude and phase to the 8 linear arrays, whose geometry is alternated to help cancel out any cross-polarisation, which was measured to be $-33\,$dB in the plane of the linear array and $-22\,$dB in the corporate feed plane. Sidelobes of around -13 dB are achieved with an efficiency of 58%. Beamwidths are 7·2° x 7·3° and the input VSWR is 1·37 at 9.4 GHz; at the broadside frequency the VSWR is about 4.

The circularly polarised form of the rampart line (Hall, 1979, Wood *et al.*, 1978) has been produced as a two-dimensional array for 17 GHz operation. The detailed operation of the linear rampart array is given in Sections 5.2.2 and 7.4.2. The array shown in Fig. 7.15, was fed by a triplate corporate feed, similar to that used for the comb line array described earlier. With a taper designed for $-25\,$dB sidelobes, -17 dB was achieved in the corporate feed plane. An exponential distribution was used in the other plane resulting in sidelobes of $-11\,$dB. An ellipticity of less than 3 dB was obtained over a 6% bandwidth with an efficiency of 32% for 4·5° x 4·5° beamwidths. The input VSWR was better than 2:1 over this bandwidth which included the broadside frequency.

A microstrip corporate feed was used in an array of 4 travelling wave, higher order mode arrays (Menzel, 1978) as shown in Fig. 6.13. Sidelobes of greater than $-10\,$dB were obtained in both planes for an approximately 25° x 25° beam at 10 GHz. For a 4·6 λ_m length line, about 10% of the input power was absorbed in the load. No details are given of input VSWR or efficiency.

In conclusion, this review serves to highlight the performance limitations inherent in the microstrip form. Sidelobe control is generally worse than in other

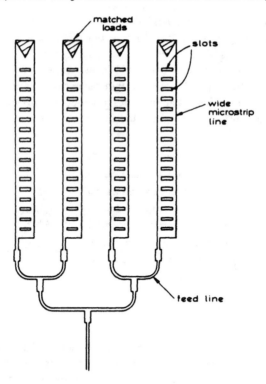

Fig. 6.13   *Microstrip higher-order mode array*

forms of antenna as also is efficiency. The reasons for these limitations are now becoming clearer and some of these were discussed in Chapter 5, in particular mutual coupling and unwanted radiation. Material and manufacturing tolerances and substrate surface waves also limit performance and these are discussed in Chapter 8. The survey of practical types suggested that few developed forms have a complete feed structure on the same substrate as the array elements and that where this is necessary radiation pattern control is limited due to unwanted feed radiation and feed array coupling (Murphy, 1979). More commonly hybrid structures are used with either the feed in microstrip or triplate behind the array or in triplate in the same plane. The problems of mutual coupling and feed design are dealt with in the following Sections.

6.3 Mutual coupling

The problem of mutual coupling between elements in linear arrays of resonant microstrip radiators was discussed in Section 5.3 and the methods presented there can be directly applied to two-dimensional forms. In particular, Bailey and Parks (1978) have shown that the scattering matrix representation of the array can be used to calculate the excitation voltages of the array element for a given impressed voltage distribution. For an $M \times N$ array, eqn. 5.8 becomes:–

$$
\begin{bmatrix} V_{11}^- \\ V_{12}^- \\ \cdot \\ \cdot \\ V_{21}^- \\ V_{22}^- \\ \cdot \\ V_{mn}^- \end{bmatrix} = \begin{bmatrix} S_{11,11} & S_{11,12} & \cdots & S_{11,21} & S_{11,22} & \cdots & S_{11,mn} \\ S_{12,11} & S_{12,12} & \cdots & S_{12,21} & S_{12,22} & \cdots & S_{12,mn} \\ \cdot & \cdot & & \cdot & & & \\ \cdot & \cdot & & \cdot & & & \\ S_{21,11} & S_{21,12} & \cdots & S_{21,21} & S_{21,22} & \cdots & S_{21,mn} \\ S_{22,11} & S_{22,12} & \cdots & S_{22,21} & S_{22,22} & \cdots & S_{22,mn} \\ \cdot & \cdot & & \cdot & & & \\ S_{mn,11} & S_{mn,12} & \cdots & S_{mn,21} & S_{mn,22} & \cdots & S_{mn,mn} \end{bmatrix} \times \begin{bmatrix} V_{11}^+ \\ V_{12}^+ \\ \cdot \\ \cdot \\ V_{21}^+ \\ V_{22}^+ \\ \cdot \\ V_{mn}^+ \end{bmatrix}
$$

$$(6.2)$$

where V^- and V^+ are the reflected and incident port voltages, respectively, and S are the scattering coefficients. For a beam scanned in the direction θ, ϕ, which are conventional spherical coordinates shown in Fig. 6.14, eqn. 5.10 for the active reflection coefficient becomes

$$
\Gamma_{mn}(\theta, \phi) = \sum_{p=1}^{M} \sum_{q=1}^{N} S_{mn,pq} \left| \frac{V_{pq}^+}{V_{mn}^+} \right| \exp[jk_0 \{(m-p)d_x \sin\theta \cos\phi
$$

$$
+ (n-q)d_y \sin\theta \sin\phi\}] \tag{6.3}
$$

where k_0 is the free space wave number and d_x and d_y are the spacings in the x and y directions, respectively. The excitation voltage on element (m, n) is then

$$
V_{mn} = V_{mn}^+ [1 - \Gamma_{mn}(\theta, \phi)] \tag{6.4}
$$

There are at present no published results on the effect of mutual coupling in planar arrays similar to those for linear arrays given in Section 5.3 although the problem is being examined. Thus it is not possible to estimate what limitations exist on array performance due to coupling. It is expected, however, that some improvement in performance will be obtained using the above methods for coupling compensation.

Section 5.4 discussed the practical performance limitations on the comb line array as a representative and well investigated example of the series-fed linear array and included some effects and calculations on mutual coupling. The conclusion on the calculation of such effects was that, although the use of a surface current approach could in principle allow simultaneous modelling of these effects, the use of a transmission line model is straightforward and to some extent can model such effects. This can be extended to comb to comb coupling and eqn. 5.25 may be used. Although this effect has not been included in the analysis, measurements of radiation patterns for various comb to comb spacing were made, (Hall and James, 1978), and these show that this coupling is significant. Table 6.2 shows the measured performance versus comb spacing of three $15 \lambda_0 \times 15 \lambda_0$ size comb line arrays at 17 GHz. This indicates that a comb spacing of less than $0 \cdot 846 \lambda_0$ is necessary to prevent unacceptable grating lobes. However the sidelobe level in the

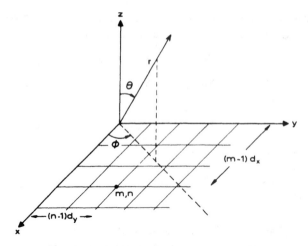

Fig. 6.14   *Array co-ordinate system*

Table 6.2   *Measured performance of two-dimensional comb line arrays with various comb spacings*

Number of combs	Comb spacing, λ_0 (centre line to centre line,)	Grating lobe level dB	Sidelobe level dB	
			H-plane	E-plane
16	0·846	−35·0	−18·2	−17·0
14	1·009	−18·5	−20·25	−11·0
12	1·130	−15·0	−20·2	−13·0

H-plane, that is in the plane of the comb line, is reduced for wider spacing indicating that mutual coupling is disturbing the aperture distribution down the comb line. The increase in *E*-plane sidelobe level is attributed to the reduction in the number of combs and hence the reduced filling in of the *E*-plane distribution. Problems within the corporate feed may also be responsible and these are discussed in the next Section.

Sidelobe levels below −23 dB have been achieved for the 15 λ_0 × 15 λ_0 comb line array. In the *H*-plane, comb performance was optimised empirically within the array environment. In the *E*-plane, feed design was improved and unwanted feed radiation reduced.

These results and those in Chapter 5 for the linear array indicate that mutual coupling is an important effect in array design but difficult to calculate for large arrays of dissimilar elements. The problem in hybrid resonator arrays, such as the disc array, is analogous to that in phased arrays of waveguide openings, where a wide literature and a good understanding of its effects exists. Thus it is expected that progress will be made in the analysis and design of arrays that are corrected for coupling effects and this will allow improved control of sidelobe levels. The problem in series arrays however does not seem to be so clear cut. Primarily because of the complicated geometry, the analysis of such arrays is much more difficult and to date only transmission line models have been used. Mutual couplings can be added to the model; however a scattering matrix representation of the complete array is not straightforward. This suggests that progress in this field will be slower and designs that compensate for mutual coupling will be largely empirical or at best a combination of empirical and theoretical approximations. As ultimate performance cannot be achieved by empirical methods alone, the lack of a complete analysis of the comb line array is seen as limiting the achievable array performance. It should be pointed out, though, that practical tolerance problems present a similar limit to realisable performance and may reduce any benefits brought by improved theoretical methods. It is probable that this may also apply to resonator arrays when these are produced at higher frequencies than 10 GHz where tolerance problems become more significant.

The problems of unwanted radiation and cross-polarisation in relation to linear arrays were examined in Chapter 5 and these factors will also be significant in two-dimensional arrays. In patch arrays, cross-polarisation is confined to the 45° planes and will be suppressed to a low level by the multiplicative effect of the principal plane array factors. Unwanted feed radiation will thus form the main problem and this is discussed in the next Section. In series fed arrays, and in particular the comb line, cross-polarisation occurs in the principal planes, with levels of −23 dB being measured for a $15 \lambda_0 \times 15 \lambda_0$ array by Hall and James (1978). Higher levels have been measured on smaller arrays suggesting that this may form a fundamental limit to the sidelobe suppression that can be achieved.

6.4 Techniques and problems in feed design

6.4.1 Feed design and component characterisation

Feeds for arrays that are coplanar with the microstrip array or mounted behind it can in general be made in waveguide, coaxial line, triplate, microstrip or any other suitable transmission medium. Printed structures, that are compatible with microstrip, such as triplate or microstrip are now widely used and these result in thin planar structures. Thus feed design for microstrip arrays involves either open microstrip for integrated feed-array antennas, or shielded microstrip or triplate for hybrid types. In general, the use of substrates and frequencies that produce efficient radiation from the antenna elements imply difficulties in design, due to feed radiation, if open microstrip is used. If shielded microstrip is used radiation is eliminated but the calculation of the discontinuity characteristics is rendered inaccurate due to the power lost to box modes or in any absorbent material used to suppress such modes and coupling between lines may also cause problems. In triplate these problems are largely eliminated, although parallel-plate modes may be generated at line inhomogeneities.

The problem of design of feeds for series arrays was discussed in Sections 5.2 and 5.4, and this present Section is devoted to corporate feeds that can be used in conjunction with series or corporately fed linear arrays to form two-dimensional arrays.

The basic element in a corporate feed network, which is simply a device for distributing power between n output ports with a given distribution, is an *n*-way power splitter (Nagai, 1974), or a combination of *m*-way power splitters, where *m* $< n$. This was described in Section 5.1.2 and with similar reasoning a choice of $m = 2$ appears to be a near optimum arrangement for two-dimensional arrays. The choice of the 2-way $m = 2$ splitter also minimises the ratio of maximum to minimum impedance required in the feed structure.

In such a multipath circuit, mismatches are important. If the elements connected to the outputs are mismatched then power will be reflected back into the feed. If the outputs of each two way splitter are isolated then this reflected power will only appear at the input and there will no be disturbance of the output distribution. Isolation can be achieved by means of the Wilkinson splitter, (Matthaei *et al.*, 1964), which uses an isolating resistor, or a branch line coupler, (Gupta, 1976), shown in Fig. 6.15(a) and (b). Both of these elements have the disadvantage of having a limited bandwidth, the Wilkinson element having the added disadvantage of using an add on component which reduces the basic simplicity of the printed construction of the microstrip array. Feeds using branch line couplers, which need a matched load on one port, have been made in microstrip at *S* band (2 to 4 GHz) with monopole radiators, having a −28 dB sidelobe level, (Gupta, 1976). Although such splitters can be broadband, (Matthei *et al.*, 1964), the overall size of each splitter is increased and in addition little data on their design is available at higher frequencies.

A wide bandwidth non-isolating power splitter can be formed from a T-junction or an inline splitter as shown in Fig. 6.15(c) and (d). Such elements have been used for some time, (Sommer, 1955), and have been shown to have wide bandwidth characteristics, although the outputs are not isolated. The approximate design requires that

$$\frac{1}{Z_{in}} = \frac{1}{Z_1} + \frac{1}{Z_2} \tag{6.5}$$

where Z_{in}, Z_1 and Z_2 are the input and output line impedances, respectively. Design of the corporate feed then requires the correct selection of the power split ratio in each splitter to give the required output distribution. Application of such design methods to triplate or microstrip requires the calculation of the line width to give the desired impedance in each medium. In asymmetric splitters, phase errors arise at the bends and tapers, Fig. 6.15(e), that can be compensated for by altering the arm lengths as shown in Fig. 6.15(f). Use of a triplate corporate feed, in addition to reducing radiation losses, avoids the need for compensation for the line width dependent phase constant of microstrip lines.

The analysis of discontinuities in microstrip was discussed in Section 2.4, where

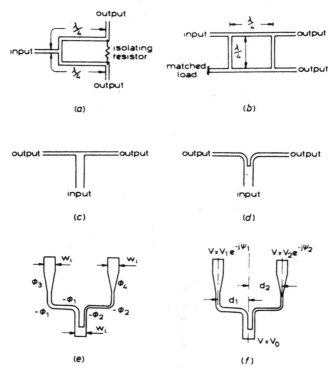

Fig. 6.15 *Types of power splitter and phase errors that occur in inline splitters*
a Wilkinson isolated splitter
b Two-arm branch line isolated splitter
c T-junction splitter
d In-line splitter
e Phase errors in an asymmetric power splitter
   ϕ = (actual electrical length − physical mean path length)
   ϕ_1, ϕ_2 = phase errors at bends, ϕ_3, ϕ_4 = phase errors at tapers
   w_i = line width at input and outputs
f Compensation method for phase errors in (e)
   If $\Psi_1 > \Psi_2$ then $d_1 > d_2$ for $\exp(-j\omega t)$ time variation

it was noted that none of the available methods take account of radiation and are thus limited in accuracy when applied to integral feeds array structures. Little data is available on discontinuities in shielded microstrip; in practice the design data for open microstrip is used and no problems are reported (Cipolla, 1979).

The analysis of discontinuities in the triplate line poses similar problems to that of microstrip in that the field form of the basic mode cannot be written down in closed form. An approximation which allows the problem to be solved by various methods is to assume magnetic walls at the edge of the line which now has an equivalent width w_e. The latter is found by equating the impedance of the closed line thus formed to the impedance of the triplate line. This is shown diagrammatically in Fig. 6.16. Although Fig. 6.16(b) has the triplate impedance, as it

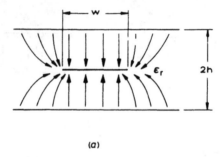

(a)

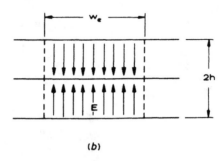

(b)

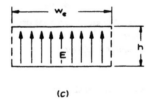

(c)

Fig. 6.16   *Derivation of the triplate equivalent waveguide model*
   a Triplate structure
   b Section of parallel plate transmission line of equal impedance as triplate in (a)
   c Equivalent waveguide model

consists of two halves that are isolated, it is sufficient to consider one half, Fig. 6.16(c), where w_e is approximately given by (Gunston, 1972):

$$w_e = 2h\pi \ln\left[\frac{2(1+\sqrt{k})}{1-\sqrt{k}}\right]^{-1} \quad \text{for} \quad \frac{w}{2h} < 0.5 \tag{6.6}$$

$$= \frac{2h}{\pi} \ln\left[\frac{2(1+\sqrt{k_1})}{1-\sqrt{k_1}}\right] \quad \text{for} \quad \frac{w}{2h} > 0.5 \tag{6.7}$$

where

$$k = \text{sech}\left(\frac{\pi w}{4h}\right) \tag{6.8}$$

and

$$k_1 = \tanh\left(\frac{\pi w}{4h}\right) \tag{6.9}$$

The fields in the closed waveguide can be found for the various modes that can exist. The equivalent circuit of a discontinuity may then be found by integral equation methods (Marcuvitz, 1951) or mode matching techniques, (Wolff *et al.*, 1972). The solution of the integral equations is considerably simplified if low frequency operation is assumed while mode matching gives the frequency dependence and is especially useful for wide equivalent waveguides where overmoding is possible. Both methods have been used for waveguide discontinuities, (Marcuvitz, 1951, Kuhn 1973); integral equation methods are found to give results that are accurate to within 5 or 10% at low frequencies while the mode matching technique gives good accuracy at higher frequencies. Expressions for the frequency dependence of equivalent circuit of an impedance step have been given by (Schwinger and Saxon, 1968) and give results for the series inductance that agree well with results using the mode matching method up to the onset of the first higher-order mode.

Results using integral equation methods for discontinuities in waveguide can be transformed to the triplate equivalent waveguide, (Altschuler and Oliner, 1960) and a typical result for an impedance step is shown in Fig. 6.17(a).

The series reactance, X, of such a step is given by

$$\frac{X}{Z_1} = \frac{2w_{e1}K}{\lambda_t} \tag{6.10}$$

and the reference plane extensions by:

$$l_1 = -l_2 = 2h\frac{\ln 2}{\pi} \tag{6.11}$$

The low-frequency approximation (Altschuler and Oliner, 1960) gives:

$$K = \ln\left[\text{cosec}\left(\frac{\pi\alpha}{2}\right)\right] \tag{6.12}$$

where

$$\alpha = \frac{w_{e2}}{w_{e1}}$$

and the frequency-dependent solution (Schwinger and Saxon, 1968) is:

$$K = \ln\left(\frac{1-\alpha^2}{4\alpha}\right)\left(\frac{1+\alpha}{1-\alpha}\right)^{1/2\,(\alpha+1/\alpha)} + 2\left(\frac{A+A'+2C}{AA'-C^2}\right)$$

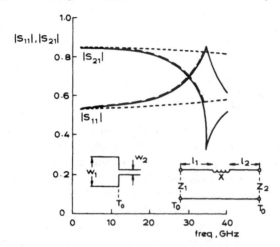

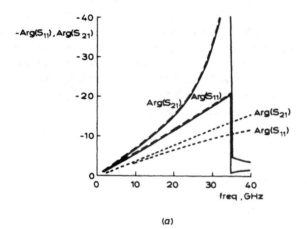

(a)

Fig. 6.17   *S-parameters of triplate discontinuities*
$2h = 1 \cdot 59 \, \text{mm}, \, \epsilon_r = 2 \cdot 32, \, w_1 = 5 \, \text{mm}, \, w_2 = 1 \, \text{mm}$
——— mode matching, 16 modes (Hall and James, 1981)
– – – eqn. 6.13
· · · · · eqn. 6.12
a Impedance step

$$+ \left(\frac{w_{e1}}{4\lambda_t} \right)^2 \left(\frac{1-\alpha}{1+\alpha} \right)^{4\alpha} \left[\frac{5\alpha^2 - 1}{1 - \alpha^2} + \frac{4\alpha^2 C}{3A} \right]^2 \qquad (6.13)$$

where

$$A = \left(\frac{1+\alpha}{1-\alpha} \right)^{2\alpha} \frac{1 + (1 - w_{e1}^2/\lambda_t^2)^{1/2}}{1 - (1 - w_{e1}^2/\lambda_t^2)^{1/2}} - \frac{1 + 3\alpha^2}{1 - \alpha^2}$$

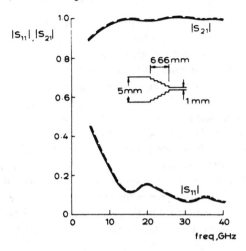

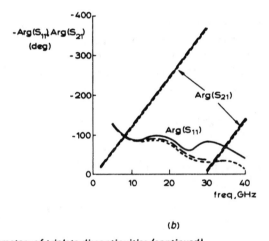

(b)

Fig. 6.17   *S-parameters of triplate discontinuities (continued)*
$2h = 1 \cdot 59$ mm, $\epsilon_r = 2 \cdot 32$, $w_1 = 5$ mm, $w_2 = 1$ mm
——— mode matching, 16 modes (Hall and James, 1981)
– – – eqn. 6.13
· · · · · eqn. 6.12
b Linear taper, length = $6 \cdot 66$ mm, number of steps = 16

$$A' = \left(\frac{1 + \alpha}{1 - \alpha}\right)^{2/\alpha} \frac{1 + (1 - w_{e2}^2/\lambda_t^2)^{1/2}}{1 - (1 - w_{e2}^2/\lambda_t^2)^{1/2}} + \frac{3 + \alpha^2}{1 - \alpha^2}$$

$$C = \left(\frac{4\alpha}{1 - \alpha^2}\right)^2$$

and λ_t is the wavelength in the triplate line. Alternatively, the mode-matching technique can be applied (Hall and James, 1979, 1981), and the result for the same step is also shown in Fig. 6.17(a). Using this method, the equivalent circuit parameters are not directly calculated; rather the reflection and transmission coefficients ρ_n and τ_n for each mode, n, are calculated taking into account the effect of cutoff modes and the S parameters found from ρ_0 and τ_0 for the fundamental TEM mode. Fig. 6.17(a) shows that the modal method predicts discontinuities in the S parameters at 34·5 GHz, the cutoff frequency for the first symmetrical higher-order mode. These are not observed using eqn. 6.12, which only agree with the modal results and eqn. 6.13 at low frequencies. Eqn. 6.13 is only valid up to this cutoff frequency. Fig. 6.17(b) shows the S parameters of a typical linear taper. The taper is split into many noninteracting small steps which are then cascaded together. As each step is small there will be little coupling to the higher-order mode. It can be seen that all three methods agree well and that the device operates smoothly through the higher-order mode cutoff frequency. This indicates that a corporate feed made up of smooth tapers and bends will have a wider bandwidth than one composed of steps and sharp bends. It is also concluded that for smooth components, such as the taper, that either method will give adequate theoretical characterisation.

For more complex discontinuities, such as the T-junction and in-line splitter the mode matching method may well prove to be the most convenient approach. Hall and James (1979, 1981) have applied mode-matching to the inline splitter shown in Fig. 6.18 and examined the effect of load mismatches and errors in the output line widths and the results throw some light on the possible performance limits of

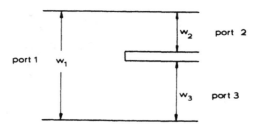

Fig. 6.18   *Inline power splitter geometry*

simple corporate feeds using such non-isolating power splitters. Table 6.3(i) shows the effect of placing loads of reflection coefficient, ρ, at arbitrary distances from the junction. Equal mismatches produce negligible effect on S_{21} and S_{31}. However, asymmetrical loads or differences in the distance to the loads created significant changes in the amplitude of S_{21}, S_{31} accompanied by differential phase shifts. Similarly errors in the output line widths, Table 6.3(ii), of the order of the etching tolerance also produce differential phase and amplitude errors. The junction is particularly sensitive to errors in the narrow output line which suggests that feeds producing particularly deep tapers will be very sensitive to manufacturing errors. Hall and James suggest that for a 16-output corporate feed designed for a −30 dB side-lobe level limit, using $2h = 1·59$ mm at 17 GHz, tolerances may lead to a lower

Table 6.3   *Effect of load conditions and tolerances on output line widths on an asymmetrical in-line power splitter.*
$w_1 = 2.571$ (mm), $w_2 = 0.199$ (mm), $w_3 = 1.742$ (mm), $2h = 1.59$ (mm), $\epsilon_r = 2.32$, freq $= 17 \, GHz$

(i) Load conditions

	ρ_2	arg (ρ_2)	ρ_3	arg (ρ_3)	S_{21}	S_{31}	arg (S_{21}) − arg (S_{31}) deg
1 Matched splitter	0.0	0.0	0.0	0.0	0.508	0.86	0.0
2 Equal mismatches	0.2	0.0	0.2	0.0	0.509	0.86	0.0
3 Asymmetrical mismatch	0.0	0.0	0.2	0.0	0.569	0.827	−6·4
4 Phase errors in mismatches	0.2	5.0	0.2	0.0	0.507	0.863	−1·1

(ii) Tolerances

	w_2 mm	w_3 mm	S_{21}	S_{31}	arg (S_{21}) − arg (S_{31}) deg
1 Matched output lines	0.199	1.742	0.509	0.861	0.0
2 Oversize w_2	0.30	1.742	0.530	0.847	−2.55
3 Undersize w_2	0.10	1.742	0.468	0.877	2.27
4 Oversize w_3	0.199	1.842	0.501	0.865	0.43
5 Undersize w_3	0.199	1.642	0.516	0.856	0.57

limit to the achievable phase front control of about ±4°. Using empirical optimisation ±7° has been achieved in practice.

6.4.2 Transitions between triplate and coaxial lines and microstrip

The problem of exciting an open waveguiding structure is not a new one and the presence of radiation from the feed transition has long been appreciated as an important and somewhat uncontrollable factor when good radiation pattern control is required.

In microstrip arrays, various feed types are used. In integral feed-array antennas one or more transitions from coaxial line to the feed network is used. In resonator arrays fed from the rear, one or more coaxial feed-throughs are used for each element. In hybrid arrays fed from a coplanar network the triplate to microstrip junction forms the feed point. Thus the coaxial to microstrip line or resonator and the triplate to microstrip line junctions form the primary feed discontinuities.

Neither of these discontinuities are amenable to simple analysis for similiar reasons to those given in Section 2.4 Chapter 2. However, estimates have been made of the radiation loss and some methods of radiation reduction suggest themselves. Lewin (1960) has treated the coaxial feed-through probe shown in Fig. 6.19 (a) in which the coaxial line is assumed to be matched to the microstrip. Using a single filamentary strip current plus dielectric polarisation current model of the microstrip

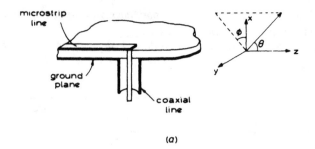

(a)

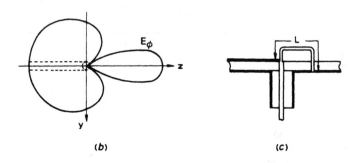

(b)   (c)

Fig. 6.19   *Matched coax to microstrip transition*
    a Cross-section of junction and co-ordinate system
    b Radiation pattern in $\phi = 90°$ plane
    c Cross-section of short-circuit stub

line, discussed in detail in Chapter 3, the effect of the current in the post together with its image in the ground plane is to produce an electric far-field,

$$E_\theta = -\frac{je^{-jk_0r}60k_0h}{r\sqrt{\epsilon_e}} \cos \phi \qquad (6.14)$$

$$E_\phi = \frac{je^{-jk_0r}60k_0h}{r\sqrt{\epsilon_e}} \left[\frac{\epsilon_e - 1}{\sqrt{\epsilon_e} - \cos \theta} - \sqrt{\epsilon_e} \right] \sin \phi \qquad (6.15)$$

and the radiated power, P, is given by

$$P = 60(k_0h)^2 \left[1 - \frac{\epsilon_e - 1}{2\sqrt{\epsilon_e}} \ln \left\{ \frac{\sqrt{\epsilon_e} + 1}{\sqrt{\epsilon_e} - 1} \right\} \right] \qquad (6.16)$$

where k_0 is the free space wave number and h the substrate height. The form of the radiated field is shown in Fig. 6.19(b). For a dielectric constant, $\epsilon_e = 2·25$, P reduces to

$$P = 0·33 \{60(k_0h)^2\} \qquad (6.17)$$

which is about one third of that radiated by an open circuit when calculated in the same way and with an equal current wave incident on it. Lewin suggests that P may be substantially reduced if a short-circuited stub is used as shown in Fig. 6.19(c). If the overall length of the stub perimeter, L, is given by

$$L = \frac{8hZ_s}{Z_m\sqrt{\epsilon_e}} \tag{6.18}$$

where Z_s and Z_m are the stub and microstrip line impedance, respectively, then P is reduced to

$$P = 0.07 \{60(k_0h)^2\} \tag{6.19}$$

or to about one-fifth that of the uncompensated transition.

The coaxial feed-through connected to a microstrip resonant radiator such as the disc or half-wavelength patch will be much less a source of unwanted radiation as it is part of the radiating element itself and will only give rise to small distortions in the element radiation pattern, which, in turn, will not greatly influence the array patterns. The problem of modelling resonator feeds is discussed further in Chapter 9.

Henderson and James (1981) have examined both coplanar coaxial to microstrip and triplate to microstrip transitions using a variational technique that employs a continuous eigenvalue mode representation of the fields. The method is an extension of that used in Chapter 3 (James and Henderson, 1979) to calculate the equivalent circuit of the open circuit microstrip line. Fig. 6.20(a) shows the configurations analysed. Fig. 6.20(b) shows the percentage of the power transmitted by the transition that is lost to radiation for the triplate to microstrip case. This is compared to an estimation (Lewin, 1975) for a coaxial feed through transition and it is seen that the loss is similar. The power radiated by a coplanar coaxial launcher of various diameters is shown in Fig. 6.20(c). Here it is seen that the power lost can be reduced by reducing the diameter of the coaxial line while maintaining a matched transition. Any power radiated from the feed region will affect the sidelobe level of the antenna array and Fig. 6.20(d) shows a pessimistic estimate of the sidelobe level S_L which could be contributed by these unwanted radiation sources. The solid curve indicates the level, S_L, contributed by the scattering of surface waves generated in the substrate calculated using the method of James and Henderson (1979), Sections 3.4 and 3.5, which again is a pessimistic estimate. It can be seen that this level is of the same order as those levels obtained by feed radiation. Although small reductions in power radiated by these transitions can be made by adjusting the geometry as indicated, it is evident that there is little scope for drastically reducing radiation loss from feed transitions. Radiation pattern perturbations may however be reduced by covering the transitions with absorbent material (Hall and James, 1981).

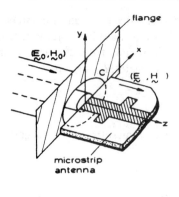

(a)

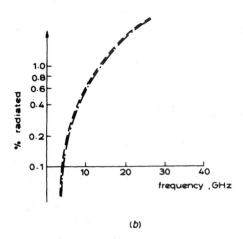

(b)

Fig. 6.20   *Coax and triplate to microstrip transitions*
a General model of feed transition
b % power radiated by a 50Ω line to 50Ω polyguide microstrip feed
—.—.— triplate line, $h = 0.79$ mm, $\epsilon_r = 2.32$
····· coax probe launcher of the type shown in Fig. 6.19(a)

6.5 Array gain limitations due to losses

The conduction losses in microstrip feed line forms a significant limit to the achievable gain in microstrip arrays. As the array size is increased the directional gain will increase proportionately. However, as array size increases so the feed line lengths become longer and the feeder loss will eventually increase faster than the directional gain and the power gain will therefore decrease. The value of the power gain maximum will depend on the feeding method employed; for example, centre-fed series arrays will have lower losses and hence a higher maximum than end fed arrays.

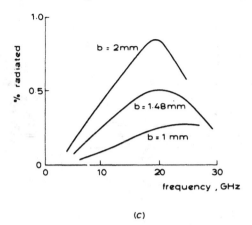

(c)

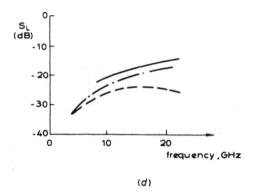

(d)

Fig. 6.20   *Coax and triplate to microstrip transition (continued)*
 c % power radiated by 50Ω coaxial launchers of outer radius *b* onto 50Ω alumina
 lines, *h* = 0·5 mm
 d Pessimistic sidelobe level S_L for a polyguide antenna array, *h* = 0·79 mm due to
 feed radiation and surface wave generation
 ———— surface wave
 —.—.— triplate
 · · · · · · coaxial launcher

. or corporate fed arrays Collier (1977) has quantified the limit. The feed geometry is shown in Fig. 6.21(a), where each line feeds a ground plane slot and there
are $2L/\lambda_0$ and L/λ_0 elements in the *E*- and *H*-planes, respectively, where L^2 is the
area of the square array. The directional gain, *D*, (Bach, 1972) is thus:

$$D = G_{el} + 10 \log_{10} \frac{2L^2}{\lambda_0^2} \quad \text{(dB)} \tag{6.20}$$

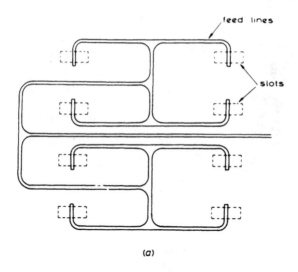

(a)

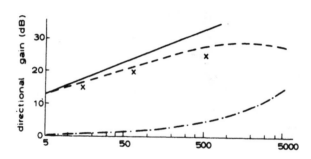

(b)

Fig. 6.21   *Microstrip ground plane slot array losses*
 a Array corporate feed arrangement
 b Power gain of array against number of elements
 —— theoretical gain of array with no feeder losses
 ‑ ‑ ‑ ‑ ‑ theoretical gain of array with feeder losses
 —.—.— feeder attenuation
    x   measured values of gain

where G_{el} is the element gain in decibels. For the geometry shown the length of feeder line from the input point to any element is $3L/2$. Thus the power gain of the array is given by:

$$G = G_{el} + 10 \log_{10} \frac{2L^2}{\lambda_0^2} - \frac{3L}{2} F \quad \text{(dB)} \qquad (6.21)$$

where F is the feeder loss in dB/unit length. To find the maximum of this expression let $dG/dL = 0$ which yields:

$$L = \frac{13.3 \log_{10} e}{F} = \frac{5.776}{F} \tag{6.22}$$

Thus for Collier's array at 12 GHz where $F = 0.075$ dB/cm the maximum gain can be seen from Fig. 6.21(b) to be about 30 dB. The lower measured gain is attributed to extra losses in the splitters and bends.

For two-dimensional comb line array the maximum power gain (Hall and James, 1978) is estimated to be about 28 dB, based on two measured values and one calculated assuming a line loss of 0·1 dB/microstrip wavelength, Fig. 6.22. This agrees well with that calculated using eqns. 6.21 and 6.22 as the average feeder length is comparable to $3L/2$.

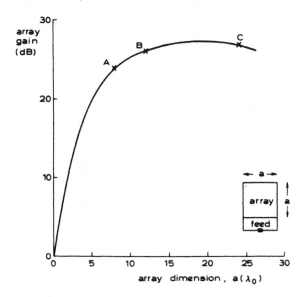

Fig. 6.22   *Gain limit for a microstrip end-fed travelling-wave two-dimensional array*
    A, B measured values, C estimate based on 0·1 dB/λ_m.

A particular solution to the loss problem, for the case of corporately fed arrays, is to split the array into several sections and feed each section by a corporate feed in a lower loss medium, such as coaxial line. This approach is used in an synthetic aperture radar antenna for the SEASAT satellite, (Murphy, 1979). The antenna is split into 8 panels to allow it to be folded to fit the launch vehicle shroud. Each panel consists of 8 × 16 rectangular patch elements and this is then split into eight 4 × 4 element arrays fed by a corporate feed in 0·6 cm diameter coax as shown in Fig. 6.23. The losses in the feed network external to the array was measured to be 0·6 dB and a power gain of 34·9 dB was achieved.

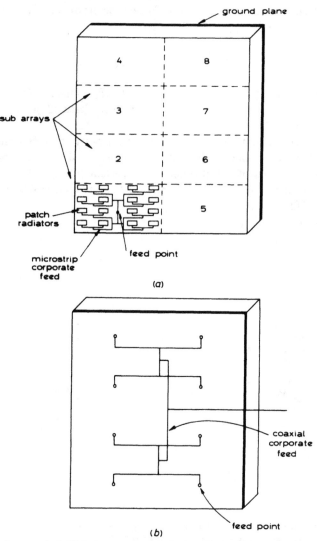

Fig. 6.23   *One panel of 1024-element microstrip array for SEASAT satellite antenna*
     a Front view, showing 8 subarrays of microstrip patches with integral microstrip
     corporate feeds
     b Rear view of one panel showing coaxial corporate feed connnecting 8 microstrip
     subarrays

6.6 Summary comments

This Chapter reviews the wide variety of forms of two-dimensional array that are
available to meet many varied applications. The fundamental performance limitations of all of them are not precisely known yet but detailed studies of some

forms such as the patch antenna and the comb line array, do reveal limiting factors. These are manufacturing and material tolerances, considered again in Chapter 8, mutual coupling, unwanted co- and cross-polarised radiation, transition radiation and conduction losses. Mutual coupling affects the radiation patterns and input impedance and can be particularly troublesome in scanned arrays. It was apparent that its effect needs to be understood and quantified and the design adjusted accordingly. Here to some extent the principles established for waveguide scanning arrays can be brought in. However, the problems do not yet appear to be severe, in particular no blind spots have been noticed, possibly because of the small size of the arrays so far built and tolerance effects. Coupling between the feed and array in integrated feed array antennas is a particularly difficult problem and may be an important limitation for this type of arrangment; little scope may be available for its reduction and hybrid solutions must be sought. Unwanted radiation from the feed is also a problem, with cross-polarised lobes far out from the copolarised main beam becoming significant in small comb line arrays. The design of the feed itself is also important, in maintaining the required array distribution. Transitions between the feed and array also appear to be a fundamental limitation of hybrid forms, although posing less of a problem than an integrated feed. Whether the transition can be placed at the centre of the array and its radiation incorporated as part of the aperture distribution is a topological problem worthy of further consideration. Little is known about feeds to resonant patches and the limits are not yet clear. Lastly, line losses are significantly higher than in other media and place a limit on the maximum gain. The losses, largely in the conductor, may not be significantly reduced without disturbing the constructional simplicity of the microstrip form. Nevertheless, the gain limit can be overcome by linking smaller arrays together with a lowloss cable. However, in spite of all these problems the available forms of microstrip arrays do exhibit considerable advantages in size, weight and cost over existing types and the attraction of trading these for ultimate performance is becoming apparent.

6.7 References

ALTSCHULER, H. M., and OLINER, A. A. (1960): 'Discontinuities in the centre conductor of symmetric strip transmission line', *IRE Trans.*, MTT-8, pp. 328–339

BACH, H. (1972): 'Directivity diagrams for uniform linear arrays', *Microwave J.*, **15**, pp. 41–44

BAILEY, M. C., and PARKS, F. G. (1978): 'Design of microstrip disc antenna arrays'. NASA Technical Memorandum 78631, Langley Research Centre, Hampton, VA, USA

BYRON, E. V. (1972): 'A new flush mounted antenna element' in 'Phased array antennas', Proc. 1970 Phased Array Antenna Symposium, pp. 187–192

CASHEN, E. R., FROST, R., and YOUNG, D. E. (1970): 'Improvements relating to aerial arrangements'. British Provisional Patent (EMI Ltd.) Specification 1294024

CIPOLLA, F. W. (1979): 'A 7·5 GHz microstrip phased array for aircraft satellite communications'. Proc. workshop on Printed Antenna Technology, New Mexico State University, USA, pp. 19–1 to 19–18

COLLIER, M. (1977): 'Microstrip antenna array for 12 GHz TV', *Microwave J.*, **20**, pp. 67–71

DERNERYD, A. G. (1976): 'Linearly polarised microstrip antennas', *IEEE Trans.*, **AP-24**, pp. 846–851

EASTER, B. (1975): 'Equivalent circuits of some microstrip discontinuities', *IEEE Trans.*, **MTT-23**, pp. 655–660

EMI-VARIAN (1973): 'Printed antennae and front ends', *Microwave J.*, **18**, pp. 20F–20G

GUNSTON, M. A. R. (1972): 'Microwave transmission-line impedance data' (Van Nostrand Reinhold Company, London)

GUPTA, C. B. (1976): 'Build an integrated dolph-Chebyshev array', *Microwaves*, Nov, pp. 54–58

HALL, P. S. (1979): 'Rampart microstrip line antennas'. European Patent Application 79301340.0

HALL, P. S., and JAMES, J. R. (1978): 'Microstrip array antennas'. Final Report on Research Agreement AT/2160/033RL, Royal Military College of Science, Shrivenham, UK

HALL, P. S., and JAMES, J. R. (1979): 'Analysis and design of triplate corporate feeds at high frequency'. Proc. 9th European Microwave Conference, Brighton, pp. 106–110

HALL, P. S., and JAMES, J. R. (1981): 'Design of microstrip antenna feeds, Part 2: design and performance limitations of triplate corporate feeds', *IEE Proc.*, Pt.H, **128**, pp. 26–34

HALL, P. S., GARRETT, C., and JAMES, J. R. (1978): 'Feasibility of designing millimetre microstrip antenna arrays'. Proc. AGARD Conf. 245 on Millimetre and Submillimetre Wave Propagation & Circuits, Munich, pp. 31–1 to 31–9

HENDERSON, A., and JAMES, J. R. (1981): 'Design of microstrip antenna feeds, Part 1: estimation of radiation loss and design implications', *IEE Proc.*, Pt.H, **128**, pp. 19–25

HENRIKSSON, J., MARKUS, K., and TIURI, M. (1979): 'A circularly polarised travelling-wave chain antenna', Proc. 9th European Microwave Conference, Brighton, pp. 174–178

JAMES, J. R., and HALL, P. S. (1977): 'Microstrip antennas and arrays, Part 2: new design technique', *IEE J.*, MOA, **1**, pp. 175–181

JAMES, J. R., and HENDERSON, A. (1979): 'High frequency behaviour of microstrip open-circuit termination', *IEE J.*, MOA, **3**, pp. 205–218

KUHN, E. (1973): 'A mode-matching method for solving field problems in waveguide and resonant circuits', *Arch. Elek. Ubertranung*, **27**, pp. 511–518

LEWIN, L. (1960): 'Radiation from discontinuities in strip-line', *Proc. IEE.*, **107C**, Monograph 358E, pp. 163–170

LEWIN, L. (1975): 'Theory of waveguides' (Newness-Butterworth, London)

MARCUVITZ, N. (1951): 'The waveguide handbook', (McGraw-Hill, New York)

MARKOPOULOS, D., and CATECHI, P (1977): 'Directive microstrip antennas'. Proc. 7th European Microwave Conf., Copenhagen, pp. 288–291

MATTAEI, G. L., YOUNG, L., and JONES, E. M. T. (1964): 'Microwave filters, impedance matching networks and coupling structures' (McGraw-Hill, New York)

MENZEL, W. (1978): 'A new travelling wave antenna in microstrip'. Proc. 8th European Microwave Conf., Paris, pp. 302–306

METZLER, T. (1979): 'Microstrip series arrays'. Proc. Workshop on printed circuit antenna technology, New Mexico State University, USA, pp. 20–1 to 20–16

MUNSON, R. E. (1974): 'Conformal microstrip antennas and microstrip phased arrays', *IEEE Trans.*, **AP-22**, pp. 74–78

MURPHY, L. R. (1979): 'SEASAT and SIR-A microstrip antennas'. Proc. workshop on printed antenna technology, New Mexico State University, USA, pp. 18–1 to 18–20

NAGAI, N. (1974): 'Basic considerations on TEM-mode hybrid power dividers', *Bull. Res. Inst. Appl. Electr. (Japan)*, **26**, pp. 25–41

NISHIMURA, S. NAKANO, K., and MAKIMOTO, T. (1979): 'Franklin-type microstrip line antenna'. Int. Symposium Digest, IEEE Antennas and Propagation Soc., Vol 1, Seattle, Washington, pp. 134–137

OLTMAN, G. H. (1978): 'Electromagnetically coupled microstrip dipole antenna elements'. Proc. 8th European Microwave Conf. Paris, pp. 281–285

SOMMER, D. J. (1955): 'Slot array employing photo-etched triplate transmission line', *IRE Trans.*, **MTT-3**, March, pp. 157–162

SCHWINGER, J., and SAXON, D. S. (1968): 'Discontinuities in waveguides' (McGraw-Hill, New York)

TIURI, M., HENRIKSSON, J., and TALLQUIST, S. (1976): 'Printed circuit radio link antenna' Proc. 6th European Microwave Conf. Rome, pp. 280–282

WHEELER, H. A. (1964): 'Transmission properties of parallel wide strips by a conformal mapping approximation', *IEEE Trans.*, **MTT-12**, pp. 280–289

WHEELER, H. A. (1965): 'Transmission line properties of parallel strips separated by a dielectric sheet', *IEE Trans.*, **MTT-13**, pp. 172–185

WILLIAMS, J. C. (1977): 'Cross-fed printed aerials'. Proc. 7th European Microwave Conf., Copenhagen, pp. 292–296

WILLIAMS, J. C. (1978): 'A 36 GHz printed planar array', *Electron. Lett.*, **14**, pp. 136–137

WOLFF, I., KOMPA, G., and MEHRAN, R. (1972): 'Calculation method for microstrip discontinuities and T-junctions', *Electron. Lett.*, **8**, pp. 177–179

WOOD, C., HALL, P. S., and JAMES, J. R. (1978): 'Design of wideband circularly polarised microstrip antennas and arrays'. IEE Conf. on Antennas & Propagation, London. IEE Conf. Publ. 169, Pt. 1, pp. 312–316

YEE, J. S., and FURLONG, W. J. (1979): 'An extremely lightweight fuselage-integrated phase array for airborne applications'. Proc. workshop on printed circuit antenna technology, New Mexico State University, USA, pp. 15–1 to 15–12

Chapter 7

Circular polarisation techniques

This Chapter is concerned with the various techniques which may be used to generate circularly polarised radiation from microstrip antennas. Circular polarisation is particularly useful for a number of radar, communication and navigation system requirements because of the behaviour it exhibits upon reflection from regular objects of reversing the hand of circularity to produce predominantly orthogonal polarisation. The system will then tend to discriminate against reception of such reflected signals in favour of direct paths or reflections from irregular shapes. A radar system using circular polarisation will therefore receive reflected signals from irregularly shaped aircraft rather than nearly spherical raindrops, or a communication system will be able to reject multipath reflections from ground or large buildings. A further attribute is that in a communication system using circular polarised radiation the rotational orientation of the transmitter and the receiver are unimportant in relation to the received signal strength, whereas with linearly polarised signals there will be only very weak reception if the transmitter and receiver orientations are nearly orthogonal. An example of this is a telemetry link to a small aircraft or missile which is capable of performing complex manouvers. The available techniques of generating circular polarised radiation from antennas is of great interest then for many systems applications.

In the discussions on the radiation mechanism of microstrip antennas given at various points in this book the main reliance has been made on relating radiation to magnetic current sources derived from the electric field at the edge of the antenna conductor pattern. The local radiation polarisation is thus essentially linear with the electric field normal to the edge of the conductor pattern, the dual of the situation found with current-radiating wire antennas. Local generation of circular polarisation would require the colocation of electric and magnetic current sources with appropriate phasing, a situation that is impossible to produce with a pure microstrip antenna configuration although it has been achieved in the form of a printed circuit slot/dipole antenna with a shallow backing cavity (Sidford, 1973 and 1975). The construction and mode of operation of that element are, however, not closely related to the microstrip antenna concept (see Section 1.3.1).

Since the 'natural' radiation of a microstrip antenna source is linearly polarised, generation of circular polarisation can only be achieved by combining individual sources with appropriate orientations and phasings. This may be done in a number of ways such as by feeding separate orthogonal modes of one conductor pattern such as the square or circular patch as has been briefly discussed in Chapter 4, or by radiating from small sources formed by suitable discontinuities in a travelling wave transmission line as in the rampart line antenna mentioned in Chapter 5. The currently available methods of generating circular polarisation thus break down into what are essentially two 'families' of antennas, orthogonal resonators and travelling-wave system, and in the following Sections the treatment will be in that order.

7.1 Orthogonal resonators with separate feeds

The first form of circular polarised microstrip antenna to be considered is one where the radiators consist of orthogonally polarised elements which may either be individual conductor patterns or isolated modes in a single symmetric structure, fed by a power-splitting network with appropriate amplitudes and phases. The main flexibility in design of these antennas comes from the possible methods of feeding the orthogonal modes. Two families of feed networks are readily identified, these being the reactive types providing no isolation between the output ports and the coupler types requiring built-in loads which give isolation. The relative merits of these two types are discussed in Section 7.1.2 in relation to their performance as a function of the element mismatch variation with frequency. Whichever type of feed network is used, its design will be influenced by the choice of either making it a separate unit linked to the antenna by coaxial lines or integrating it on the same printed circuitboard as the antenna. In the former case the position of the feed point may be chosen so as to produce the required input match as discussed in Section 4.4.1, but if integration is chosen a very high impedance (see Table 4.2) will be presented to the feed line which will make the incorporation of transformer sections necessary.

7.1.1 Polar diagram control

Numerous practical examples of circular polarised resonant microstrip antennas have been developed, but one important constraint which should normally be met for single circularly polarised elements is that the phase centres of the radiation from the two orthogonally polarised sources be at the same point. If this is not the case, the circular polarisation would degrade as a function of angle of observation due to the changing path lengths to the phase centres. Antenna configurations such as that of Fig. 7.1(a) have therefore not been used in practice owing to the resulting limitations of the polar diagram. The use of square or circular patches as circularly polarised antennas had been discussed by Brain and Mark (1973), Munson (1974), Howell (1975) (Fig. 7.1(b)) and some polar diagrams were presented in Chapter 4 for these two cases which showed that there was little difference between the results obtained. Combinations of quarter-wave short-circuit patches have also been used to give circular polarised signals by Sanford and Klein (1978) as shown in Fig. 7.1(c), and other arrangements may be used as in Fig. 1.1. These short-circuit patch

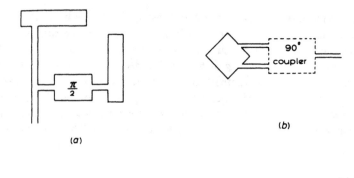

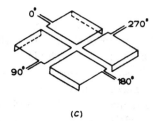

Fig. 7.1   *Some possible configurations of microstrip patches for circular polarised applications*
    a Offset rectangular open-circuit patches
    b Dual feed to square open-circuit patch
    c 'Crossed-slot' using four short-circuit patches

arrays have normally been utilised when a requirement exists for a nominally circular polarised antenna which maintains signal coverage down to angles close to the ground plane, although the radiation is, of course, forced to be linearly polarised at that point by the action of the groundplane in suppressing the parallel polarised component.

The control of polar diagrams available is quite limited and can only be achieved by changing from one form of element to another. The basic open-circuited patch antenna has different pattern shapes in the E- and H-planes, and this leads to circular polarisation degradation at angles away from broadside even if perfect circularity is produced at broadside. Since the H-plane pattern is principally controlled by the $\cos \theta$ multiplying terms (eqns, 4.26 d and 4.33 b), pattern equalisation can only be obtained by narrowing the E-plane pattern by increasing the spacing between the interferometer radiating source (Fig. 7.2) (a)). The dimensions of the open-circuit patches are fixed by the resonance requirement, so such flexibility is only possible by using a shorted-patch configuration as in Fig. 7.2(b).

7.1.2 Effect of radiator mismatch

In most cases the main frequency sensitive parameter in the antenna/feed network system will be the resonant characteristic of the antenna element, since even the simplest forms of feed network give reasonable performance over bandwidth of 10–20% provided they operate with constant load parameters. The choice of feed

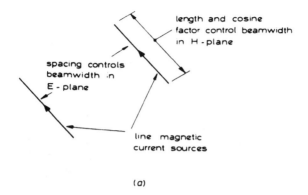

length and cosine
factor control beamwidth
in H -plane

spacing controls
beamwidth in
E - plane

line magnetic
current sources

(a)

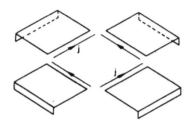

(b)

Fig. 7.2   *Control of beamwidth in E- and H-planes of magnetic current source pair*
    a Relation between dimensions and beamwidths
    b Implementation using short-circuit patches

network is limited to hybrid types such as the branch line or ratrace of Figs. 7.3(a) and (b), or T-splitters of either matched or unmatched form as in Figs. 7.3(c) and (d). The last three units inherently provide cophased output signals and therefore require an excess quarter wavelength of line in one output arm to generate the required 90° phase relation at the antenna elements. It is important to know the way in which each of these feed network types behave as a function of the match presented by the antenna elements in order to understand trade-offs between the simplicity of some forms and the better performance of others. The essential division is between the first three types which incorporate a fourth port with an absorbing load and are designed to give good isolation between the output ports and the unmatched T-splitter which is purely reactive and thus does not isolate the output ports. Theoretical calculations by Wood and James (1980) show that the following characteristics are found.

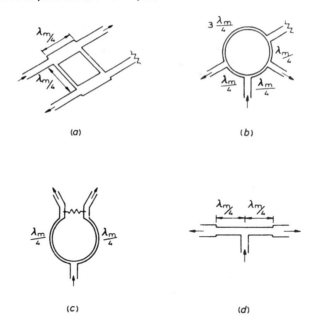

Fig. 7.3   *Feed networks giving equal power split into two channels*
   a Branch line hybrid
   b Ratrace network
   c Matched T-network (Wilkinson splitter)
   d Reactive T-network

(a) Isolated configurations (Figs. 7.3 (a) (b) and (c)):
(i) Because of the 90° phase shift between the output arms, the signals reflected by the antenna elements are returned to the absorbing load and the overall input match remains good over the band.
(ii) Because of the isolation between the output ports, signals reflected back into the network are not retransmitted into the outputs, and the output amplitude and phase split therefore remains good.
(iii) The overall effect upon the system performance is therefore identified as being just loss of gain due to the VSWR of the antenna elements, which at 2:1 is of the order of $\sim$0·5 dB.

(b) Reactive T-splitters (Fig. 7.3(d)):
(i) Because of the 90° phase shift between the ouput arms, the reflection from the antenna elements tend to cancel at the input port so that the input match remains acceptable ($< \sim 1.2$:1) over the band.
(ii) The lack of isolation between the output arms is inherently accompanied by a mismatch between the output port and the splitter, so that reflections from one antenna element will be retransmitted into both output arms. The 90° phase shift in one arm will however mean that the total output signal in each arm will be different, so that the amplitude and phase split will be degraded. Calculations on a

wide variety of T-splitter configurations show that all generate > 3 dB ellipticity when the element VSWR exceeds about $1 \cdot 4:1$.

(iii) There are no power losses directly in the power splitter and the overall input match remains good, thus all the power incident on the unit is radiated. This does not mean, however, that there is no gain loss since the degradation of the ellipticity is due to power being radiated in the unwanted hand of circular polarisation. The ellipticity expressed as a voltage ratio is identical to the VSWR of the linearly polarised element, and this means that the gain loss of the wanted hand of circular polarisation with a reactive feed will be the same as that with isolated output feed networks. The power absorbed in the internal load of the isolated output devices is therefore equivalent to the power radiated in the unwanted hand of circular polarisation by the reactive T-splitter.

It should of course be remembered that these conclusions are drawn from studies assuming that perfect networks are used and in practice errors will occur in manufacture which degrade the performance. The dominant effect in the case of the reactive T-splitter will remain the susceptability to the antenna elements' mismatch, whilst in the case of the isolated output networks the errors introduced in manufacture will be more significant. A further point to note is that an antenna system incorporating an isolated output network will be less influenced by environmental aspects such as temperature change since this would show up mainly as changes in the VSWR response of the antenna elements as discussed in Chapter 8. Finally, the advantage of the reactive T-splitter is that it does not require the built-in load which may be difficult to provide, particularly where the feed network is placed on the antenna element printed circuit board. It is shown in Section 7.2 that a somewhat simpler structure is available which gives identical performance to the reactive T-network.

Examples of antennas which have been built using matched and unmatched T-splitters are shown in Figs. 1.1 and 7.4; measured results of ellipticity and input match which illustrate the above conclusions are given in Fig. 7.5.

7.1.3 Cross-fed patch arrays

In the design of circularly polarised patch array antennas, the previously described techniques of ensuring that each element is circularly polarised may be employed together with a separate system for distributing power to the array elements. The use of patch elements capable of supporting two modes which generate the required orthogonally polarised radiation separately does allow a configuration to be employed which simplifies the feed network arrangements (Derneryd, 1976 a). This antenna configuration, shown in Fig. 7.6, was derived from the line array formed by coupling a number of square patch elements in series by means of half-wavelengths of microstrip transmission line as discussed in Section 5.2.1. If the array is made two-dimensional, this method of feeding may be used to excite orthogonal polarisations and the total radiation may be made circular polarised by feeding the two input ports to the orthogonal polarisations from a power splitter with appropriate phasing. The total path lengths to the first array element in each channel

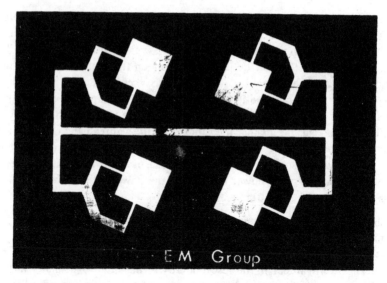

Fig. 7.4   *Photograph of circularly polarised array of square patches with offset reactive T-net-work feeds*
Substrate $h = 1.59$ mm, $\epsilon_r = 2.32$, frequency = 10.4-11 GHz

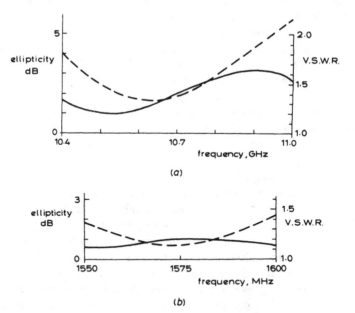

(a)

(b)

Fig. 7.5   *Performance of circular polarised antennas using matched and reactive T-networks*
———   VSWR
- - - - -   Axial ellipticity
a Reactive T-network feeding square patches (Fig. 7.4) Patch bandwidth for VSWR
of 2:1 ~610 MHz
b Matched T-network feeding four short-circuit patches (Fig. 1.1) Patch bandwidth
for VSWR of 2:1 ~ 28 MHz

should be kept as near equality as possible, however, to avoid degradation of the ellipticity by phase mistracking. The symmetrical excitation at one port was found to be necessary to ensure an isolation between the two modes of $-20\,dB$. Results presented by Derneryd for a 4×4 element array designed to operate at $9\,GHz$ showed that the polar diagram beam widths and gains of the two polarisations were virtually identical, however, no information was given on the performance as a function of frequency. It is not, however, expected that this arrangement would be capable of operating satisfactorily over wide bandwidths since the line array configuration itself has a narrower bandwidth for 2:1 VSWR than does an isolated patch (Derneryd, 1976 b), and the direct coupling of the individual square elements to each other may lead to some degradation of the circularity as the VSWR of the elements varies with frequency.

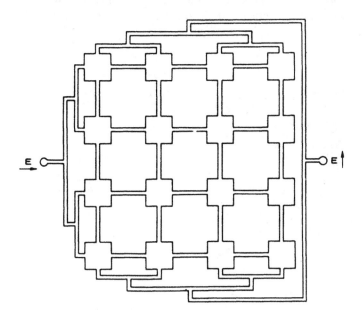

Fig. 7.6   *Cross-feeding of array of square patches to generate circular polarisation*

7.2 Resonators with single feed points

In situations where space is at a premium both on the antenna circuit board and at the rear of the board, the incorporation of a feed network to excite the orthogonal modes of a square or circular patch antenna may be very difficult. It would thus be useful if an antenna could be driven so as to radiate circular polarisation with only a single feed position. In the case of the square antenna, owing to the fact that the frequency responses of the orthogonal modes are identical the modes are always excited in phase or in antiphase by a single feed point, with only the relative amplitudes being controlled. For the circular disc units, complete rotational symmetry means that excitation by a single feed point just gives rise to a single mode with its peak voltages lying on the same radius line as the feed. In each case, therefore, only linear polarisation may be generated with these symmetrical configurations. The key to obtaining circular polarisation with a single feed point lies in the detuning of the two modes by slightly changing the shape of the patch or by including capacitive or inductive elements in the structure, and the reasons for this will be discussed in the following section. Many configurations have been proposed in the literature, Sanford and Munson (1975) describing a rectangular patch whereas Ostwald and Garvin (1975) and Kerr (1978, 1979) propose the use of tabs and slots respectively to achieve a circular polarised configuration (Fig. 7.7). Although these configurations are quite different from each other, they all operate by the same method of slightly detuning the degenerate modes of the symmetrical structure as noted above. The basic characteristics of all forms are therefore the same and can be outlined by a fairly simple model as follows.

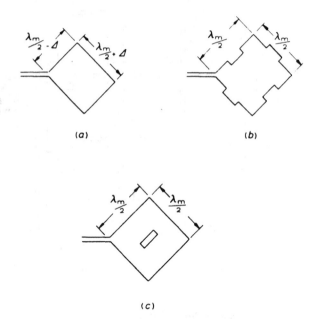

(a)   (b)

(c)

Fig. 7.7   *Circular polarisation of patches obtained by detuning of orthogonal modes*
   a Almost square patch
   b Modes capacitively detuned with tabs and gaps at edges
   c Modes inductively detuned with central slot

7.2.1 Equivalent circuit
In order to discuss the equivalent circuit of a patch antenna with a slightly detuned pair of modes, it is useful to refer back to the 'cavity mode' expansion of the fields under a patch as discussed in Section 4.5.1. The general form for the electric field is

$$E_z = \sum_m \frac{A_m}{\epsilon_r k_0^2 - k_m^2} \ \Psi_m \qquad (7.1)$$

where m is the index of the mode field structure Ψ_m, and k_m is the complex resonant wave number of the mode. If frequencies close to the resonances of two slightly detuned modes are considered, the field will be dominated by those modes:

$$E_z \sim \frac{A_{m1}\Psi_{m1}}{(\epsilon_r k_0^2 - k_{m1}^2)} + \frac{A_{m2}\Psi_{m2}}{(\epsilon_r k_0^2 - k_{m2}^2)} \qquad (7.2)$$

It is possible to analyse the conditions required to generate circular polarisation directly from eqns. 7.1 and 7.2. This requires considering the behaviour of the ratio of A_{m1} and A_{m2} as a function of frequency and the locations of the complex poles k_{m1} and k_{m2}, as in Richards *et al.* (1979); but a simpler understanding may be obtained by using an equivalent circuit analysis as follows.

The total voltage between the patch and the ground plane is the sum of the voltages due to each individual mode

$$V_{in} = V_{m1} + V_{m2} \qquad (7.3)$$

Since the modes are each represented by a parallel tuned circuit, the detuned mode-pair has the equivalent circuit of a series combination of parallel resonant circuits, as shown in Fig. 7.8.

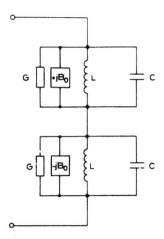

Fig. 7.8 *Equivalent circuit representation of patch with two nearly degenerate modes*

If the probe is positioned correctly within the patch, the modes are coupled equally and the equivalent circuits for the modes will be virtually identical. In the case of a square or circular patch, this location will be along the diagonals between the mode axes. If it is further assumed that the modes are detuned by equal amounts in opposite senses, the expression for the admittance of each mode may approximately be expressed by

$$Y_1 = G + j(B + B_0) \tag{7.4a}$$

$$Y_2 = G + j(B - B_0) \tag{7.4b}$$

where $B = (\omega c - 1/\omega L)$ and $\pm jB_0$ are the susceptances which detune the modes.

It can be easily shown that the ratio between the voltages across each mode radiation conductance (which is equal to the ratio of the radiated orthogonal fields) is

$$\frac{V_1}{V_2} = \frac{G + j(B - B_0)}{G + j(B + B_0)}$$

$$= \frac{G^2 + (B - B_0)^2}{G^2 + (B + B_0)^2} \left\{ \exp\left[j \tan^{-1}\left(\frac{-2B_0 G}{B^2 + G^2 - 2B_0} \right) \right] \right\} \tag{7.5}$$

For circular polarisation to be obtained, the result $V_1/V_2 = \pm j$ is required, i.e. when $G^2 + (B - B_0)^2 = G^2 + (B + B_0)^2$

and

$$G^2 + B^2 - B_0^2 = 0$$

giving

$$B = 0 \tag{7.6a}$$

and

$$B_0 = G \tag{7.6b}$$

Thus the required offset frequency for each mode, Δf_0, is related to the detuned resonant frequency f_r by the standard relationships between unloaded Q_0, frequency and circuit parameters as

$$\Delta f_0 = \frac{f_r}{2Q_0} \tag{7.7}$$

The input admittance of the equivalent circuit is given by

$$Y_{in} = \frac{G^2 - B^2 + B_0^2 + 2jGB}{2G + 2jB} \tag{7.8}$$

which at resonance $(B = 0)$ and with the circular polarisation condition $(B_0 = G)$ applied, reduces to $Y_{in} = G$.

7.2.2 Frequency response

The characteristics of this type of antenna as a function of frequency may be calculated from eqns. 7.5 and 7.8 by varying B, and the results are given in Fig. 7.9 as plots of VSWR and ellipticity. This behaviour can be compared with that of the symmetrical patch with separate feeds to the orthogonal modes discussed earlier in Section 7.1.2. It is found that the behaviour of the detuned mode circular polarised antenna is virtually identical to that of the symmetrical patch fed from a reactive

T-network, with the ellipticity degrading rapidly with frequency whilst the input VSWR deteriorates but remains within acceptable limits. Since the resulting antenna is a more compact structure, it will in general be the practically prefered configuration of the two.

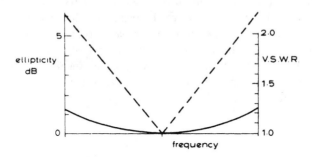

Fig. 7.9   *Theoretical performance of detuned-mode patch over frequency band corresponding to 2:1 VSWR of each mode*
- - - - -   axial ellipticity
———   VSWR

It must of course be remembered in interpreting the results of this analysis that it is very approximate in nature. For instance, the effects of mutual coupling between the two modes which may affect the equivalent circuit parameters, particularly the conductance values, has been neglected. Similarly, the behaviour of the susceptance components of the two equivalent circuits will not exactly track each other as assumed in the analysis since slight differences between the geometries of the two modes are required to give the detuning between the modes. The circuit analysis is therefore not suitable for a final performance prediction but it is adequate as a guide for experimental development.

The polar diagrams of this antenna are not controllable in any way since the dimensions are fixed by the operating requirements. At the central frequency the mode excitations are the same as those for the hybrid-fed symmetrical structure and the polar diagrams will be as shown in Fig. 4.12 or 4.16 as appropriate. The patterns will vary somewhat with frequency as the relative amplitude and phase of the modes change; this will however principally show as changes in the width of the ellipticity envelope, with the pattern shape for the nominal hand of circular polarisation being resonably stable.

7.2.3 Pentagon patch

An alternative microstrip patch antenna configuration giving circular polarisation with a single feed point has been described by Weinschel (1975) and Weinschel and Carver (1976). This antenna is in the form of a pentagon which is symmetrical about one axis (Fig. 7.10), the dimensions of the pentagon and the location of the feed point at the periphery of the patch being chosen experimentally to produce circular polarisation. No theoretical explanation of the operation of this antenna was given, and indeed in view of the quite complicated geometry involved a modal analysis using the magnetic wall model would clearly be extremely difficult. There is, however, little doubt that the principle of operation is the same as that outlined earlier in this Section in that the dimensions are chosen so as to produce the appropriate resonant frequency relationship between two modes whose field structures are similar to the fundamental modes of the regular patches. The performance and characteristics are therefore expected to be similar to those outlined above. It may in fact be concluded that there will be a wide variety of patch shapes which are capable of producing similar characteristics, however theoretical design will be virtually impossible and the nearly regular types discussed previously in this Section will be prefered practically.

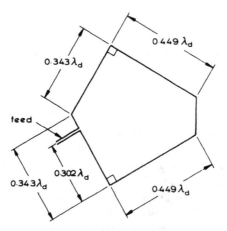

Fig. 7.10   *Dimensions of pentagon patch giving circular polarisation,* $\lambda_d = \lambda_0/\sqrt{\epsilon_r}$

7.3 Curved line antennas

The major problem that occurs with microstrip antennas is their typically narrow VSWR bandwidth which is due to the use of the radiation mechanism of open-circuited lines or patches. This carries over into the performance of circular polarised patch antennas in the form of limitations on the achievable ellipticity; an exception is the dual-feed patch with an isolated output feed network and only a slight drop in efficiency occurs. This latter configuration requires the provision of the additional feed circuitry, and will ultimately be limited in bandwidth by the tolerance of this circuitry to very high load VSWRs. This will in turn depend mainly upon manufacturing tolerances. A group of microstrip antennas exist which avoid use of the open-circuit radiation mechanism by radiating along the length of a curved microstrip transmission line, and this has led to the development of a compact element inherently producing good circular polarisation over very wide bandwidths in microstrip antenna terms (Wood *et al.*, 1978; Wood, 1979). The antenna takes the form either of a sector of a circular transmission line or one turn of a loosely wound spiral transmission line (Fig. 7.11) with a matched termination and relies for the generation of circular polarisation on the fact that a circular travelling wave current distribution with 2π radians phase progression per turn inherently radiates circular polarisation along the axis of the circle (Knudson, 1953).

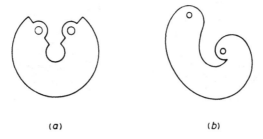

(a)   (b)

Fig. 7.11   *Circular polarised curved line antennas*
  a Sector of circular transmission line
  b Spiral transmission line

The wire-antenna version of this concept is the multiturn planar spiral antenna which has been the subject of exhaustive investigation owing to its property of operating over multioctave frequency bandwidth (Curtis, 1960). It may therefore be thought that multiturn microstrip spiral antennas would have corresponding frequency coverage, but it turns out that the radiation mechanism of the microstrip spiral is completely different and is much less effective than that of the open planar wire spiral. It is then found that as the operating frequency is increased, the outer turns on a microstrip spiral generate significantly more radiation than do those of the wire spiral, leading to serious corruption of the polar diagrams. The antenna is therefore limited in size to single turns of about one wavelength circumference. The fact that significant radiation can occur from the outer turns of such antennas indicates that there is a high proportion of the input power left at the end of the radiating curved line, and to assess this theoretically a model of the radiation mechanism will be developed in the following Section.

7.3.1 Radiation mechanism of curved line
A long straight microstrip line structure carrying the quasi-TEM mode has a balanced asymmetric surface field structure with slow-wave phase properties which gives rise to radiation only at its ends (Fig. 7.12(a), (b)) as discussed in Chapter 3. However, if the line is curved the balance is disturbed and radiation becomes possible, the characteristics of which can be identified on the basis of effective magnetic current analysis as described by Wood (1979). If it is assumed that the curvature of the line does not seriously disturb the magnitude of the surface fringe electric field at each side of the line, the total radiation from the line may be calculated from the

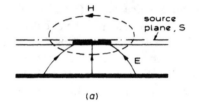

(a)

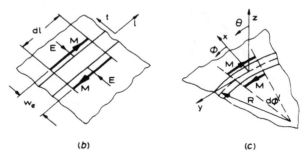

(b)   (c)

Fig. 7.12   Radiation mechanism of curved microstrip line
   a Field distribution of line in transverse plane
   b Magnetic current sources at edges of straight line
   c Imbalance in magnetic currents caused by curvature

contributions of the magnetic currents shown in Fig. 7.12(c). These two line sources are used to simplify the representation of the total effect of the complex surface distribution either side of the line, with the assumption that the presence of the backing ground-plane causes the tangential electric field to decrease rapidly away from the transmission line edge. The choice of location of the two sources is arbitrary in the absence of any detailed information of the nature of the surface field structure, but the value chosen of $w_e/2$ either side of the strip centreline has been shown to give good results in other analyses as discussed in Chapter 3. The equivalent width of the line is given by (eqn. 2.16)

$$w_e = \frac{120\pi h}{Z_m\sqrt{\epsilon_e}} \tag{7.9}$$

The radiation is then due to the fact that the geometric distortion introduced by the transmission line curvature causes the magnetic current element at the outer edge to be longer than that at the inner edge of a length of the line dl. The magnetic radiation vector contributed by this length of line is then

$$d\mathbf{L} = \mathbf{L}\, dl\, V\left(\frac{w_e}{R} + jk_0 w_e \cos\phi \sin\theta\right) \tag{7.10}$$

where V is the voltage of the line at the segment dl.

An effective magnetic current M_e may therefore be defined which acts along the centreline of the microstrip, and this is given by the expression

$$M_e = V k_0 w_e \left(\frac{1}{k_0 R} + j \cos \phi \sin \theta \right) \tag{7.11}$$

This current is therefore composed of two parts, the first being due to the curvature of the line and being independent of the angle of observation. The second, however, is due to the interferometer action of the sources at each side of the line and is thus dependent upon the angle of observation. The radiated power patterns of these two components are given by the expressions

$$P_r (1/k_0 R) \sim 1 - \sin^2 \phi \sin^2 \theta \tag{7.11a}$$

$$P_r (j \cos \phi \sin \theta) \sim (1 - \sin^2 \phi \sin^2 \theta) \cos^2 \phi \sin^2 \theta \tag{7.11b}$$

It is the first $(1/k_0 R)$ component which is of most interest for the antenna configuration being considered since radiation of circular polarisation along the axis of the curved line is required.

An estimate of the radiation from a curved line may be made by calculating the power radiated by a full circle of the same curvature, and assuming that this power is radiated equally from the circumference a decay coefficient may be derived

$$\gamma = \frac{Z_m P_r}{2 \pi R V^2} \tag{7.12}$$

where P_r is the total power radiated, R is the radius of curvature and γ the decay coefficient of the line. Evaluation of P_r is quite lengthy although straightforward, and leads to the expression

$$P_r = \frac{2 k_0 R w_e^2 V^2 \pi^2}{\lambda_0^2 Z_0} F_\beta(k_0 R) \tag{7.13}$$

where $F_\beta(k_0 R)$ is a double-series summation in $k_0 R$ which involves the propagation constant of the microstrip line through $\beta = \sqrt{\epsilon_e} \, k_0 R$. The detailed expansion of this function is given by Wood (1979) and will not be reproduced here, however, plots of F_β are given in Fig. 7.13 for various values of ϵ_e. The decay coefficient may then be derived by combining eqns. 7.12 and 7.13 to give

$$\gamma = \frac{k_0 Z_m}{Z_0} \left(\frac{w_e}{\lambda_0} \right)^2 F_\beta(k_0 R) \tag{7.14}$$

Thus the variation of γ as a function of the radius of curvature of the line is directly proportional to F_β, and therefore reaches a maximum at the same time as F_β. It can be shown that the peak value of each F_β curve as shown in Fig. 7.13 occurs at values of $k_0 R$ which correspond approximately to the condition of 2π radians phase progression per turn. If it is assumed that a 50Ω microstrip line with $\epsilon_e = 2$ is being considered, eqn. 7.13 may be used to give $w_e = \lambda_0/4$ for an efficiency of 50% (total loss of 3 dB). This value is quite possible to achieve, for instance a 50Ω line on a substrate with $\epsilon_r \sim 2 \cdot 3$ and $h \sim 1 \cdot 6$ mm has an equivalent width of $\sim 0 \cdot 28$ λ at 10 GHz. The result does confirm the comment made earlier that this micro-

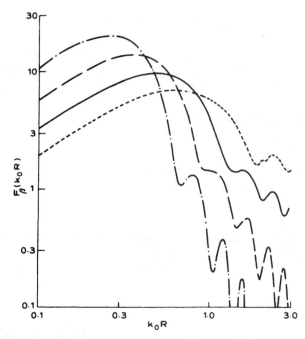

Fig. 7.13   *Curvature factor $F_\beta (k_0 R)$ for curved microstrip line radiation (after Wood, 1979)*

    · · · · · $\epsilon_r = 1$
    ——— $\epsilon_r = 2$
    — — — $\epsilon_r = 4$
    —·—·— $\epsilon_r = 8$

strip configuration must be confined to single-turn devices since the loss is not sufficient to avoid strong radiation from outer turns where the phase progression per
turn is more than 2π radians. In comparison, a well designed planar wire spiral
would be expected to have a loss through the 2π radians per turn zone of the order
of 20 dB.

The expression given above in eqn. 7.14 for the decay coefficient must be
treated with some care if it is desired to apply it to a more general curve, because it
is strictly only valid for curves which comprise a full circle. The difficulty which
arises is essentially that of interaction between radiation from different parts of the
curve, which will be the strongest for the 2π radians condition when the radiation
on axis dominates. For curves with phase progression much greater than 2π radians
per turn this effect will be less pronounced and the assumption of a continual
'throwing-off' of radiation as the wave progresses will be reasonable, so for that case
the decay coefficient expression will also give a useful value. The expression should
not, however, be applied to short sections of line whose radius of curvature is small
in terms of wavelength.

7.3.2 *Characteristics of antennas*

The radiation model and decay coefficient for the curved line given above may be applied to predict the fundamental characteristics of the antenna, and some details of the dependence of the performance of both spiral and circular curves on a number of parameters was given by Wood (1979). The most sensitive characteristic was found to be the axial ellipticity, and configurations which gave good results for this were good in other respects also. One result was that avoidance of waves propagating around the curves in the reverse direction was found to be important, this would of course be expected since such a reverse wave would strongly radiate signals of the reverse-hand of circular polarisation. During the theoretical calculations a 50Ω nominal impedance was concentrated upon since most systems impedances are of that value, and it was found that line or terminating impedances within a $50 \pm 7\cdot5\Omega$ range produced less than 1 dB variation in the ellipticity envelope. This corresponds to a VSWR of $\sim 1\cdot15{:}1$, which should be achievable in practice. Of the geometric parameters studied, those giving the greatest effect were found to be the size of the gap in the circular form needed between the feed and load points and the pitch of the spiral form. Increasing the gap in the circular form gave rise to rapid deterioration in the ellipticity and it should therefore be kept as small as possible. The spiral pitch however was found to have little effect until it became greater than the maximum radius of the curve, when the ellipticity degraded rapidly and this therefore forms the constraint on this configuration.

One characteristic which the theoretical analysis predicted was the display of beam squint with frequency. This was largely unaffected by changes in parameters, and was attributed to the generation of strong components in the radiation polar diagrams corresponding to radiation from travelling wave sources with zero and 4π radians phase progression per turn. These components would have nulls in the axial direction and a phase characteristic in the azimuth plane different to that of the wanted components, thus leading to the beam-squint phenomenon which is therefore a fundamental characteristic of the antenna. This effect was found in practice, and in general the results of measurements support the theoretical conclusions very well. The performance obtained from antennas both of the circular and spiral forms are summarised in Table 7.1, with examples of the polar diagrams shown in Fig. 7.14.

The bandwidth obtained is dependent on the tolerance placed on efficiency at low frequencies and the beam-squint at high frequencies; the VSWR and ellipticity being relatively stable over wide frequency bands. Degradation of the latter two controlling factors is quite constant with increasing offset from the nominal frequency, and a useful bandwidth of up to 40% can be obtained.

It is possible to make a few comments upon the choice of substrate for this type of antenna based upon the analysis of the decay coefficient. If the restriction is made that $Z_m = 50\Omega$, and the fact that for k_0R values corresponding to 2π radians phase progression the value of F_β from Fig. 7.13 is proportional to $\sqrt{\epsilon_e}$ is noted, eqns. 7.9 and 7.14 may be combined to give

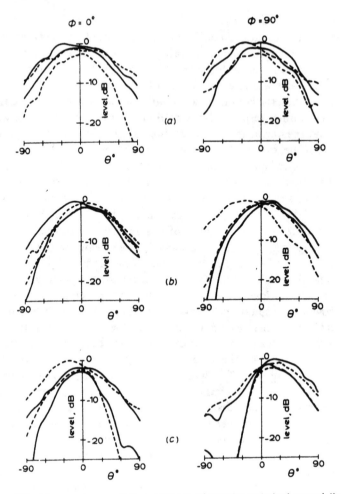

Fig. 7.14   *Ellipticity envelope radiation patterns of circular polarised curved line antennas (after Wood, 1979)*
======= circle sector
= = = = spiral curve
a 8 GHz
b 10 GHz
c 12 GHz
Envelopes show limits of signal received from rotating linearly polarised transmitter
Substrate $h = 1·59$ mm, $\epsilon_r = 2·32$

$$\gamma \sim \frac{h^2}{\sqrt{\epsilon_e}} \tag{7.15}$$

Although ϵ_e is dependent to a secondary extent upon the line geometry, it varies essentially directly with the substrate permittivity. The expressions thus shows that the efficiency will increase with increasing height and reducing substrate permittivity. However, as γ is increased in this way the linewidth must also be increased to maintain Z_m at 50Ω, and this provides an upper limit on efficiency since the line-width cannot be physically greater than twice the radius of curvature. If the approximation is made that $w_e \sim w + 2h$, which is found to be quite close in practice, the constraint on h is found to be

$$h < \frac{\lambda_0}{\pi(120\pi - 2Z_m\sqrt{\epsilon_e})} \tag{7.16}$$

For a value of $\epsilon_e = 2$ this expression gives $h_{max} \sim \lambda_0/15$. If it is noted that for the antennas whose performance is presented in Table 7.1 the substrate thickness was $\sim \lambda_0/20$, the conclusion is drawn that the efficiency of this class of antennas can only be improved to levels slightly better than those quoted above. Also, since the decay coefficient with these restrictions is proportional to h^2, the efficiency will reduce rapidly with decreasing h and the antennas are therefore restricted in application to those situations where a substrate thickness of $\sim \lambda_0/20$ is acceptable.

Table 7.1 *Summary of performance of curved line antennas; h = 1·59 mm, ϵ_r = 2·32*

Frequency	8–12 GHz
Bandwidth	40%
VSWR	< 1·3:1
Ellipticity	< 3 dB
Gain	0–6 dB
Efficiency	30–70%
Squint	< 25°

7.4 Travelling-wave arrays

As noted in Section 7.1.1, the main constraint on the design of an isolated circularly polarised antenna having a broad beam is the requirement that the phase centres of orthogonally polarised components of the radiation should be located at the same place. This arises owing to the fact that if the orthogonal components appear to originate from different locations there will be a variation in their relative phases as the angle of reception varies which will lead to degradation of the circular polarisation. The rate of degradation is a very sensitive function of position; for instance, if the orthogonal polarisation were correctly phased at broadside for perfect circularly polarised radiation a spacing of as little as $0.055\,\lambda_0$ between their phase centres would be sufficient to produce 3 dB ellipticity at 30° from broadside in the plane of the offset. There is, however, one configuration where such a separation may be tolerated, and this is when the antenna specification calls for fixed narrow beam radiation polar diagram. An array configuration may then be chosen in which the individual elements give good circular polarisation only over the region of space occupied by the main beam of the antenna. This would of course result in general in a sidelobe structure of quite random elliptically polarised radiation, but there is seldom any requirement upon polarisation purity in the sidelobes and normally just their total power will be specified as being less than a particular level.

Microstrip antennas may be composed of corporate network-fed individual radiators, but in that case the interelement spacing will be in excess of $\lambda_0/2$ and so there will be no problem normally in using one of the patch designs described earlier which have collocated phase centres for the orthogonal polarisations. An alternative method of constructing arrays is to use travelling-wave transmission line excitation of microstripline elements, a number of configurations being discussed in Chapter 5. Two of these may be designed so as to produce circularly polarised radiation on the main beam peak although the radiation essentially originates from linearly polarised sources of varying orientation spread out along the length of a microstrip transmission line. These two configurations, the rampart line (Hall, 1979; Wood *et al.*, 1978) and the chain antenna (Henriksson *et al.*, 1979) as shown in Fig. 7.15, were discussed briefly in Section 5.2.2 where it was noted that they allow some control in their radiated polarisation by variation of geometric parameters. Some further discussion will now be given individually for these two types in

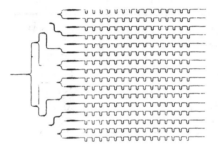

(a)

(b)

Fig. 7.15   *Travelling-wave antenna configurations giving circular polarisation*
a Rampart line antenna
b Chain antenna, after Henriksson *et al.* (1979)

terms of the constraints needed in each case for the generation of circularly polarised radiation. It is, however, worth noting at this point that while the approach taken to explain the radiation mechanism for the two forms is completely different, being magnetic current sources for the rampart line and line electric current sources for the chain antenna, this is largely a result of the limited available information on these antennas and the different backgrounds to their development. As noted elsewhere in this book, the two approaches are virtually equivalent and in most cases give theoretical predictions differing by less than the accuracy of experimental evidence.

Also, while the experimental circularly polarised chain antenna described by Henriksson *et al.* (1979) is constructed of a conductor pattern etched on a thin substrate and spaced away from the ground-plane by a very low dielectric constant foam material and is thus not a simple printed circuit antenna, the spacing involved ($\sim$0·05 λ_0) is typical of microstrip antennas and the excitation is applied between the conductor pattern and the ground-plane as with all microstrip antennas. The use of the foam spacer, however, does mean that the waves travelling along the line have rather weaker coupling than in the case of conventional microstrip construction with dielectric constants between 2 and 2·5, and since this will tend to enhance the radiation efficiency slightly some caution should be taken in directly comparing the results quoted for this antenna with other microstrip variants.

7.4.1 Rampart line antenna

The basic unit of the rampart line antenna (Hall, 1979) is a section of microstrip transmission line incorporating four mitred corners, with the input and output lines travelling in the same direction and being in line with each other , Fig. 7.16. Control of the polarisation radiated by the cell is given by the two lengths *d* and *s*, the third length *L* being fixed by the phasing between cells needed to give the required array scan angle. Wood *et al.* (1978) explain the radiation from this cell structure in terms of magnetic currents located at each edge of the line. With the quasi-TEM mode assumed on the line the sources along the straight lengths are oppositely directed and therefore tend to cancel each other's radiation, however at each corner there is an excess length of current at the outer mitred edge which acts as the dominant radiation source giving a radiated electric field polarised diagonally relative to the corner, Fig. 7.16(a). The total radiated field in the direction normal to the board is best analysed in terms of the diagonally polarised fields (E_a and E_b) as shown, and the ratio of the latter is

$$\frac{E_a}{E_b} = -\exp(-j\beta (d + s)) \tag{7.17}$$

The relative phase required to generate circular polarisation is an odd multiple of $\pi/2$, so the relationship

$$\frac{2\pi}{\lambda_m} (d + s) = (2n + 1) \pi/2 \tag{7.18}$$

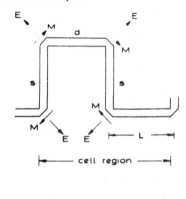

(a)

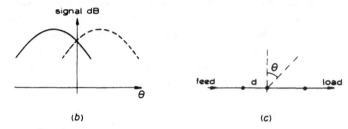

(b)   (c)

Fig. 7.16   *Radiation characteristics of cell element in rampart antenna*
a Radiation sources
b Behaviour of radiation patterns of orthogonal 45° polarisations in transverse
plane if $s \neq \lambda_m/2$
c Scanning of main beam from broadside

results between the three side lengths. The geometry is not yet completely specified
however, since the absolute values of d and s have not been fixed by these two
relationships. The final useful constraint is obtained by considering the behaviour
of the radiation polar diagrams of the two diagonal polarisations in the plane perpendicular to the array long dimension. If the value of s is chosen to be $\lambda_m/2$, then
both polarisations have their peak at broadside and the radiation patterns will be
matched to each other. However, if any other value of s is chosen, the beams will be
scanned away from broadside in opposite directions due to the transmission line
wave travelling in opposite directions along the transverse arms. Although amplitudes at broadside will remain equal, the patterns will no longer match and the
circularity of the radiation will degrade at angles away from broadside, Fig. 7.16(b).
The value of s is therefore chosen as $\lambda_m/2$ for best consistency of polarisation with
angle in this plane.

Eqn. 7.18 gives the required value of d to generate circular polarisation at broadside to the antenna board, however as noted earlier the value of L may in fact be
chosen so as to scan the main beam of the array away from the broadside. If this

is desired, then use of eqn. 7.18 would lead to degradation of the circular polarisation on the peak of the beam owing to the displacement of the orthogonal
polarised sources by the distance in the plane of the array. This effect may be
avoided by the use of a constraint on d which generates a $(2n + 1) \pi/2$ radians
phase difference between the two polarisations in the required direction. If the
beam is required to be scanned by an angle θ in the array plane away from the feed
end of the cell, Fig. 7.16(c), the modified relationship is

$$\frac{2\pi}{\lambda_m} (d + s) - \frac{2\pi}{\lambda_0} d \sin \theta = (2n + 1) \pi/2 \tag{7.19}$$

Implementation of the scan in this direction does not affect the argument leading
to the conclusion that $s = \lambda_m/2$ is required, and substituting this into eqn. 7.19
gives

$$d = \frac{(2n - 1)}{4} \frac{\lambda_m}{1 - \sin \theta \dfrac{\lambda_m}{\lambda_0}} \tag{7.20}$$

Choice of successive values of n lead to alternating hands of circular polarisation;
inversion of the pattern of the antenna conductor also leads to reversal of the hand
of circular polarisation. There is therefore no point in choosing any value of n other
than that to give the minimum value of d, i.e. $n = 1$ giving

$$d = \frac{\lambda_m}{4 \left(1 - \sin \theta \dfrac{\lambda_m}{\lambda_0}\right)} \tag{7.21}$$

Finally, the value of L may be specified by the requirement that the contribution
from successive cells must add in phase in the direction θ. This results in the equation

$$\frac{2\pi}{\lambda_m} (2s + d + L) = \frac{2\pi}{\lambda_0} (d + L) \sin \theta + 2\pi\nu$$

or

$$L = \frac{(\nu\lambda_m - 2s)}{\left(1 - \dfrac{\lambda_m}{\lambda_0} \sin \theta\right)} - d \tag{7.22}$$

where ν is the minimum integer required to give a positive value for L. Substituting
the previously derived values for the other lengths gives

$$L = \frac{3\lambda_m}{4 \left(1 - \dfrac{\lambda_m}{\lambda_0} \sin \theta\right)} \tag{7.23}$$

Final theoretical points worth noting are that the choice of lengths given above means that whilst reflections on the transmission line due to copolarised corners will add, the circular polarisation condition automatically means that the reflection from the orthogonal pairs will nearly cancel. This means that the input VSWR to the array antenna should be quite good, even for the broadside radiation condition. Also, no discussion has been given so far of the absolute radiated power level from a cell, which will clearly be a function mainly of the width of the line as with the open circuited transmission line. No data is currently available to give theoretical figures for the total power radiated, and the loss of a single cell must therefore be determined by experimental evaluation. Control of the array-plane pattern may then be achieved by varying the width of the line along the length of the antenna. It should also be noted that the radiated power can be controlled by varying s along the array. This actually controls the signal strength by scanning the pattern of the copolarised pair of corners away from broadside and may thus lead to poor side-lobes in other planes; however aperture tapering on linearly polarised arrays (Hall and James, 1981) using this method with $s < 0.06 \lambda_m$ resulted in a sidelobe level of < 20 dB with no significant lobes in other planes.

Fig. 7.17 shows results obtained for an antenna designed for a nominal 17 GHz frequency band in which no tapering has been applied to the line to control the radiation pattern. The power distribution therefore consists of an exponential taper from the feed point with an expected sidelobe level of ~ -12 dB. The results obtained bear out the general conclusion given above, and a 6% useful bandwidth was obtained based on a circular polarisation limit of 3 dB. It should be noted that the best ellipticity was only of the order of 2 dB, and the rate of degradation with frequency suggests bandwidths in excess of 10% may be achieved with an optimised design.

7.4.2 Chain antenna

The version of the chain antenna proposed by Henriksson *et al.* (1979) in a circular polarisation configuration relies for the generation of radiation upon a bent section of TEM-mode transmission line as shown in Fig. 7.18(a). The conditions for circular polarisation in a particular direction are set up by correct choice of the length s of the two arms and the bend angle α. Because α is not necessarily specified as $\pi/2$, however, the analysis of the radiation from this basic element is rather more complicated than that of the rampart line preceding. The expression for the radiated field in a general direction is given by Henriksson *et al.* (1979) for the case where the propagation constant of the microstrip line is equal to that of free space. The results may be easily modified to incorporate a different propagation constant β_m for the line, and for radiation in the plane of the array ($\phi = \pi/2$) this then reduces to the following form:

$$E_\theta \sim A \sin \frac{\alpha}{2} \cos \theta \cos \left(\frac{k_0 sT}{2} \right) \qquad (7.24a)$$

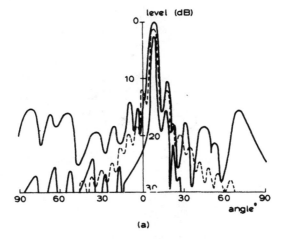

(a)

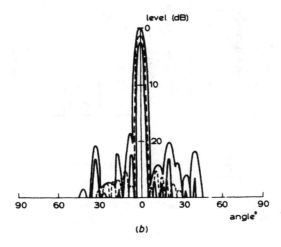

(b)

Fig. 7.17 *Ellipticity envelope radiation patterns of rampart line antenna*
———— experimental
· · · · · · theoretical
a Plane parallel to rampart line axis
b Plane transverse to rampart line axis
Envelope shows limits of signals received from rotating linearly polarised transmitter
Substrate $h = 0.795$ mm, $\epsilon_r = 2.32$, frequency = 17 GHz

$$E_\phi \sim jA \cos\frac{\alpha}{2} \sin\left(\frac{k_0 sT}{2}\right) \qquad (7.24b)$$

where

$$T = \frac{\beta_m}{k_0} - \sin\theta \sin\frac{\alpha}{2} \qquad (7.24c)$$

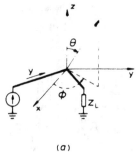

(a)

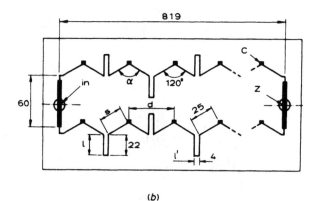

(b)

Fig. 7.18 *Analysis of radiation characteristics of chain antenna (after Henriksson et al., 1979)*
Dimensions in mm
a Circular polarised cell of array
b Array pair with Schiffmann phase shifters arranged to cancel their spurious
radiation

and

$$A = s \sin\left(\frac{\beta_m s}{2} \sin\frac{\alpha}{2} - \frac{k_0 s}{2} \sin\theta\right)\Big/\left(\frac{\beta_m s}{2} \sin\frac{\alpha}{2} - \frac{k_0 s}{2} \sin\theta\right) \quad (7.24d)$$

For circular polarisation, the only requirement is that the two components of the electric field have equal magnitude, since they are inherently in phase quadrature due to the geometry of the element. Thus

$$\tan\frac{\alpha}{2}\cos\theta = \tan\left(\frac{k_0 s T}{2}\right) \quad (7.25)$$

which does not have a unique solution for s and α, and as with the rampart line antenna an extra condition is required to fully specify the result. Henriksson *et al.* give plots of the correspondence between α and s which are required for circular

Table 7.2   *Range of chain antenna parameters*
for best performance ($\beta_m \approx k_0$)

Main beam angle, θ_m	$0° - 20°$
Length of element sections, s	$0 \cdot 25 - 0 \cdot 5\lambda$
Bend angle of element sections, α	$90° - 150°$

polarisation from eqn. 7.25. They indicate that the best practical results are obtained with values of s, α and the design main beam direction θ as given in Table 7.2 for antennas where the low relative permittivity supporting dielectric material used gives $\beta_m \sim k_0$. The choice of ranges was partly defined by their choice of minimising the beam-squint with frequency characteristic which is exhibited by all travelling-wave antennas. This in its turn was influenced by the effective length of Schiffmann phase shifters used to couple the intervening sections of the antenna, Fig. 7.18(b), whose effective length t has to be specified to fix the beam pointing angle θ_m. This effective length may be derived from the expression

$$\cos(k_0 t) = (\rho - \tan^2 (k_0 l))/(\rho + \tan^2 (k_0 l)) \tag{7.26}$$

where ρ is the ratio of the even and odd impedances of the coupled lines in the Schiffmann phase shifter. The beam pointing angle is then found from the relationship

$$\sin \theta_m = (2s + t + l' - n\lambda_0)/d \tag{7.27}$$

where n is an integer which is chosen to be unity to ensure that only one beam is generated. Again, these two relationships are obtained for the condition that $\beta_m \sim k_0$. Note that the phase shifters are arranged in image pairs Fig. 7.18(b) in order to suppress spurious radiation which would otherwise tend to be generated by the discontinuities within these devices.

Henriksson *et al.* quote results obtained both from a 3 GHz pair of antennas as shown in Fig. 7.18(b) and for a two-dimensional array at 10 GHz, and note that in order to obtain operation at broadside tuning of the individual elements to reduce reflections was required. This is due to the fact that at the broadside condition the reflections from each element add directly with each other so that large standing waves are built up causing degradation of pattern shape and ellipticity. The required tuning was achieved by means of capacitive stubs attached to the bend within each element as in Fig. 7.18(b). Array-plane polar diagram results obtained for the optimised 3 GHz case are shown in Fig. 7.19, together with the axial ratio and VSWR measured as a function of frequency. The residual power in the terminating load was about 16% at the nominal operating frequency.

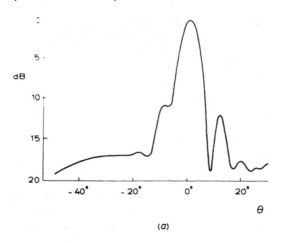

(a)

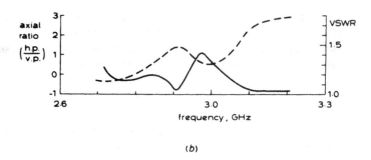

(b)

Fig. 7.19 *Performance of circular polarised chain antenna with matching tabs (after Henriksson et al., 1979)*
    a Radiation pattern in plane parallel to chain at 3 GHz
    b Performance of antenna as function of frequency
    ——— VSWR
    - - - - - ellipticity

7.5 Summary comments

Two basic design approaches to the generation of circular polarised radiation have been discussed in this Chapter, these being the use of resonant patch elements with two modes excited in the appropriate relationship or distributing sources along a suitably shaped length of microstrip transmission line. These two methods, however, lead to quite different design and operating properties due to the different radiation mechanisms involved. In the case of the resonant elements, the design characteristics are precisely defined and can be split between the radiation analysis of the patch itself as discussed in Chapter 4 and straightforward network analysis of the feed structures involved. The degree of control of the characteristics is, however, limited by the geometric constraints associated with the resonance condition, and the designs available are therefore fairly inflexible. It is not, however, difficult to obtain good performance from antennas as designed by this approach, although due to the resonant nature of the radiating element the operating bandwidth will generally be close to that of the half-wavelength open-circuited patch and will then be comparatively narrow.

The situation is quite different when the leaked radiation approach is considered, because of the avoidance of resonance effects by using unbalanced components of the distorted quasi-TEM wave produced at discontinuities as radiation sources. This then produces a transmission-line feeder with lightly loading radiators which lends itself to travelling-wave antenna concepts and their possibilities of increased bandwidth. Because the radiation sources are inherently associated with discontinuities, there are many possible configurations giving far more flexibility in configuration than the resonant approach. However, the fact that the desired sources are generally weak radiators means that unwanted radiation from other components may be a problem in some cases as with the phase shifters in the chain antenna described in Section 7.4.2. Also, theoretical analysis is far more difficult in these configurations and it is to be expected that more experimentally based development would be required, with the final operating characteristics obtained less well controlled. Overall, the travelling-wave radiators are less well advanced than the resonant elements and in view of their greater flexibility it would seem that there is scope for further invention and improvement, but the patch-based antennas will continue to provide better but more limited characteristics.

7.6 References

BRAIN, D. J., and MARK, J. R. (1973): 'The disc antenna – a possible L-band aircraft antenna'. IEE Conf. Publ. 95, Satellite Systems for Mobile Communications and Surveillance, pp. 14–16

CURTIS, W. (1960): 'Spiral antennas', *IRE Trans.*, AP-8, pp. 298–306

DERNERYD, A. G. (1976a): 'Microstrip array antenna'. Proc. 6th European Microwave Conference, Rome, pp. 339–343

DERNERYD, A. G. (1976b): 'Linearly polarised microstrip antennas', *IEEE Trans.*, AP-24, pp. 846–851

HALL, P. S. (1979): 'Rampart microstrip line antennas'. European Patent Application 79301340.0

HALL, P. S., and JAMES, J. R. (1981): 'Conformal microstrip antennas, final report on phase II, microstrip antenna for a cylindrical body application'. Research Agreement D/DRLS/5/33/11, Royal Military College of Science

HENRIKSSON, J., MARKUS, K., and TIURI, M. (1979): 'A circularly polarised chain antenna'. Proc. 9th European Microwave Conference, Brighton, pp. 174–179

HOWELL, J. Q. (1975): 'Microstrip antennas', *IEEE Trans.*, AP-23, pp. 90–93

KERR, J. (1978): 'Microstrip polarisation techniques'. Proc. antenna applications symposium, University of Illinois

KERR, J. (1979): 'Microstrip antenna developments'. Workshop on printed circuit antenna technology, New Mexico State University, pp. 3.1–3.20

KNUDSON, H. L. (1953): 'The field radiated by a ring quasi-array of an infinite number of tangential or radial dipoles', *Proc. IRE*, **41**, pp. 781–789

MUNSON, R. E. (1974): 'Conformal microstrip antennas and microstrip phased arrays', *IEEE Trans.*, **AP-22**, pp. 74–78

OSWALD, L. T. and GARVIN, C. W. (1975): 'Microstrip command and telemetry antennas for communications and technology satellites'. IEE International Conference on antennas for aircraft and spacecraft, London, pp. 217–222

RICHARDS, W. F., LO, Y. T., SIMON, D., and HARRISON, D. (1979): 'Theory and applications for microstrip antennas'. Workshop on printed circuit antennas technology, New Mexico State University, pp. 8.1–8.23

SANFORD, G. G. and MUNSON, R. E. (1975): 'Conformal VHF antenna for the Appollo-Soyuz test project'. IEE International conference on antennas for aircraft and spacecraft, London, pp. 130–135

SANFORD, G. G., and KLEIN, L. (1978): 'Recent developments in the design of conformal microstrip phased arrays'. IEE Conference on Maritime and Aeronautical Satellites for Communication and Navigation. IEE Conf. Publ. 160, London, pp. 105–108

SIDFORD, M. J. (1973): 'A radiating element giving circularly polarised radiation over a large solid angle'. IEE Conference on satellite systems for mobile communications and surveillance, London, pp. 18–25

SIDFORD, M. J. (1975): 'Performance of an L-band Aerosat antenna system for aircraft'. IEE Conference on Antennas for Aircraft and Spacecraft, London, pp. 123–129

WEINSCHELL, H. D. (1975): 'A cylindrical array of circularly polarised microstrip antennas', *IEEE Trans.*, **AP-S** (Symposium, Washington) pp. 177–180

WEINSCHELL, H. D., and CARVER, K. (1976): 'A medium gain circularly polarised microstrip UHF antenna for marine DCP communication to the GOES satellite system', *IEEE Trans.*, **AP-S** (Symposium) pp. 391–394

WOOD, C., HALL, P. S., and JAMES, J. R. (1978): 'Design of wideband circularly polarised microstrip antennas and arrays'. IEE International Conference on Antennas and propagation, London, pp. 312–316

WOOD, C. (1979): 'Curved microstrip lines as compact wideband circularly polarised antennas', *IEE J.*, MOA, **3**, pp. 5–13

WOOD, C., and JAMES, J. R. (1980): 'Study of printed circuit antennas for spacecraft radio systems'. Final report on research agreement D/DRLS/5/33/9

Chapter 8

Some manufacturing and operational problems of microstrip antennas

The salient practical problems that are likely to be encountered when the concept of the microstrip antenna is turned into reality and manufactured have been noted in the foregoing Chapters. These problems embrace both tolerance control and unwanted radiation from scattering of substrate surface waves. In this Chapter, these problems are examined and some quantitative data and estimates of performance degradation are given although there is no satisfactory way of calculating these effects to within close manufacturing limits.

There is to the authors' knowledge little published work on tolerance problems, although systems using microstrip antennas are currently in use. However, as antenna specifications become more demanding, tolerance effects are becoming evident and it is already clear that for existing substrates fundamental performance limits exist; the ultimate performance obtained may thus fall short of other types of planar antenna, such as the waveguide slot array described in Chapter 1.

The substrate surface wave problem is discussed in Chapter 3 in relation to the open-circuit end and again in Chapter 9 and there is some disagreement between the theoretical estimates of the portion of available power lost in the substrate; the calculation of the consequent performance limitations of array antennas is even more complicated and has yet to be addressed. What is clear, however, is that the analysis has revealed that surface-wave activity is a problem in low sidelobe arrays and in this Chapter we present qualitative experimental findings.

8.1 Material and manufacturing tolerance problems

The problems in microstrip antenna manufacture can be divided broadly into two categories, the characteristics of the material and the antenna production problems; some of the more important factors in these categories are given in Table 8.1. Some of these properties are given for materials commonly used in microstrip antenna production in Appendix C. It can be seen there that the quoted tolerances on the dielectric constant ϵ_r and height, h, are similar for the various materials and we may thus attribute these tolerances to quality control in manufacture rather than to

Table 8.1   *Important factors in material and manufacturing tolerance control*

Material tolerance factors	Production tolerance factors
(i) Control of substrate dielectric constant, ϵ_r, and substrate thickness, h	(i) Control of tolerances
(ii) Temperature effects	(ii) Stress relief after etching
(iii) Aging effects	

inherent properties in the material. Although no material is specifically produced for microstrip antennas all the materials quoted are 'microwave grade', which means that the dielectric loss is low and ϵ_r and h are controlled as well as possible. Material manufacturers themselves admit that they do not see tolerances being significantly reduced in the near future for reasonably priced materials due to variations in the quality of the raw materials that they use. The dielectric constant and thickness both affect antenna performance and deviations from the specified values will detune the antenna or disturb its radiation pattern or both. Temperature and aging effects cause the performance to vary with temperature and time and these effects may also be interrelated. Etching errors and stress relief after etching will affect the physical dimensions of the final antenna which in turn will lead to changes in the performance. There is little information available on these factors for low dielectric constant substrates apart from two material survey papers (Norwicki, 1979; Traut, 1979) and this means that the complete production limits on the performance cannot be fully assessed. Microstrip circuits on high dielectric constant substrates have been examined in more depth, for example Goedbloed *et al.* (1978), and this experience may be useful for antenna applications. In the next sections these problems are examined and the areas where control is as yet unacceptable or uncertain are highlighted.

8.1.1 Resonant antenna design

In small resonant microstrip antennas, such as the patch, the most critical parameter is normally the resonant frequency. Radiation control is less critical although in resonant series arrays this too may be important. Other parameters such as bandwidth and input VSWR are also less critically dependent on the practical tolerances and therefore in this Section the effect of tolerances on the resonant frequency of patch antennas are examined.

Patch antennas are designed to operate over a narrow bandwidth about their resonant frequency. For a VSWR of 2:1 and a dielectric constant of 2·32, eqn. 4.48 for the bandwidth, B, of a square patch reduces to

$$B = 81 \cdot 27 \, \frac{h}{\lambda_0} \, \% \tag{8.1}$$

where h is the substrate height and λ_0 the free space wavelength. The effect of tolerances in the substrate height and dielectric constant can be established by calculating the change in resonant frequency, f_r, given by the approximate expression

$$f_r = \frac{c}{2l\sqrt{\epsilon_e}} \qquad (8.2)$$

where c is the velocity of light, l the physical length of a half-wavelength resonator and ϵ_e is the effective dielectric constant given by eqn. 2.5, and the fringing capacitance at the resonator ends is assumed to be negligible. From this the change, Δf_r, in the resonant frequency, f_r, due to changes, δl and $\delta \epsilon_e$ in the resonator length, l, and microstrip effective dielectric constant, ϵ_e, respectively, is given by:

$$\Delta f_r = f_r \left[1 - \frac{l\sqrt{\epsilon_e}}{(l+\delta l)\sqrt{\epsilon_e + \delta \epsilon_e}} \right] \qquad (8.3)$$

The primary concern of the designer is whether the change in the resonant frequency is small with respect to the bandwidth. If this is so then repeatable system performance from a production run of antennas can be expected. To illustrate the problem, patches designed for 1 and 10 GHz on a similar substrate are examined. Table 8.2 shows that the bandwidth in MHz, calculated from eqn. 8.1, is increased

Table 8.2 *Calculated effect of tolerances in dielectric constant, ϵ_r, and substrate height, h, on resonant frequency of patches on substrate $\epsilon_r = 2\cdot32$, $h = 1\cdot59\,mm$*

Patch frequency, GHz	1·0	10·0
Bandwidth, MHz	±2·2	±217
Resonant frequency shift, Δf_r (MHz) due to		
$\delta \epsilon_r = \pm 0\cdot02\,(\pm 1\%)$	∓3·2	∓33
$\delta h = \pm 0\cdot05\,mm\,(\pm 3\%)$	∓0·2	∓11

by a factor of 100 at the higher frequency. At the lower frequency it can be seen that Δf_r calculated from eqn. 8.3 due to the manufacturers quoted tolerance on dielectric constant is larger than this bandwidth whilst at the higher frequency this shift is only about 15% of the bandwidth. The frequency shift due to the tolerance on the substrate height is considerably smaller. The conclusion is that if very thin substrates are used at low frequencies then special techniques need to be used to ensure that the resonant frequency lies within the antenna bandwidth; if not hand tuning of each antenna may be needed. Selection of dielectric sheets after careful measurement may in practice reduce the tolerances of Table 8.2 to 0·5%. Alternatively, a sample of the material may be measured and the appropriate mask chosen form a 'library' designed for various values of h and ϵ_r. It should be noted finally that in general h and ϵ_r vary by considerably less than the quoted tolerance across the board.

Changes in temperature affect the resonant frequency through thermal expansion

Table 8.3   *Calculated effect of temperature change of ±50° on resonant frequency of patches on glass-loaded PTFE substrate $\epsilon_r = 2\cdot42$, $h = 1\cdot59$ mm, dielectric constant temperature coefficient = $1\cdot4 \times 10^{-4}$ /°C, expansion coefficient = $0\cdot6 \times 10^{-4}$ /°C*

Patch frequency, GHz	1·0	10·0
Bandwidth, MHz	± 2·2	± 217
Resonant frequency shift, MHz	±9·5	±95

Table 8.4   *Effect of etching tolerance and stress relief on resonant frequencies of patches on a polyolefin substrate, $\epsilon_r = 2\cdot32$, $h = 1\cdot59$ mm. (b) and (c) from Petrie and Grove (1971)*

Patch frequency, GHz	1	10
Bandwidth, MHz	± 2·2	± 217
Resonant frequency shift, MHz due to		
(a) etching undercut = 0·36 mm, calculated	0·36	3·6
(b) stress relief after etching one face	0·3	3·0
(c) stress relief after aging 1 h at 75°C	6·2	62

and change in dielectric constant. Table 8.3 shows Δf_r, for a glass loaded PTFE substrate, for a temperature change from + 50 to − 50°C, which is not untypical for an aircraft mounted antenna. It should again be noted that these shifts are significant compared to the patch bandwidth.

Aging effects manifest themselves as movement in the substrate with repeated temperature cycling and cold flow. There is little data on the latter problem with apparent good performance maintained by many operational antennas, with some difficulties noted on more critical designs. However, stress relief due to etching and temperature cycling has been investigated by Petrie and Grove (1971). The errors in the resonant frequency due to this and etching tolerances, which are generally of the order of the copper thickness, is shown in Table 8.4. The amount of stress relief is proportional to the amount of copper etched off and for particular designs may be considerably less than shown. Similarly it will vary with the process used to apply the copper. The material can also be stabilised before etching; typically 6 cycles from −55°C to + 85°C will remove all but less than 0·1% of the total stress relief in the material.

8.1.2 Array design

All of the above factors will affect the performance of array antennas. In arrays of resonators there will, in addition, be errors in the aperture distribution caused by tolerance effects in the feed. If $f(z)$ is the aperture distribution, which will in most cases be real and $\epsilon(z)$ be the complex error on the distribution then the radiation pattern $E(\theta)$ is the Fourier transform, F of the aperture distribution:

Table 8.5   *Effect of temperature and stress relief on pointing angle of a travelling-wave array made on (1) Glass loaded PTFE substrate, $\epsilon_r = 2 \cdot 4$, (2) Polyolefin type substrate, $\epsilon_r = 2 \cdot 3$*

Error source	Change in linear dimension, $\Delta l/l$	Change in ϵ_r $\Delta \epsilon_r$	Change in pointing angle, $\Delta \theta$
(1) $\pm 50°$C temperature change	$\pm 0 \cdot 28 \times 10^{-2}$	$\mp 7 \times 10^{-2}$	$\mp 0 \cdot 06°$
(2) Stress relief			
(*a*) after etching one face	-3×10^{-4}	$-$	$+0 \cdot 02°$
(*b*) after aging 1 h at 75°C	$-6 \cdot 2 \times 10^{-3}$	$-$	$+0 \cdot 35°$

$$F(\theta) = F[f(z) + \epsilon(z)]$$

$$= F[f(z)] + F[\epsilon(z)] \tag{8.4}$$

Thus in principle the errors in the radiation pattern can be found from the distribution errors which in turn depend on the material and manufacturing tolerances. However, little information exists on the variation of the material parameters across the array substrate and this makes a more detailed analysis less useful. Hall and James, (1978) have used the design computer program for the comb line array to find the effect of changes of substrate height and dielectric constant quoted in Table 8.2 on sidelobe level and beam squint. For a $15\lambda_m$ long travelling-wave comb line, the change in the highest sidelobe level of $-16 \cdot 3$ dB, was about $0 \cdot 6$ dB and the change in beam squint was $0 \cdot 1°$. As a constant h and ϵ_r down the array was assumed this will represent an optimistic limit and for parameter changes across the board pattern changes in excess of this are to be expected.

The effect of temperature and of shrinkage due to stress relief of a travelling-wave array can be calculated from the expression

$$\sin \theta = \sqrt{\overline{\epsilon_e}} + \frac{\lambda_0}{d} \tag{8.5}$$

where θ is the pointing angle and d the element spacing. The error, $\Delta\theta$, in the pointing angle for a $\pm 50°$ temperature change at a nominal $\theta = 0°$ are given in Table 8.5 together with errors produced by stress relief (Petrie and Grove, 1971). It can be seen that there is a small change in θ with temperature and this may be significant in some applications, particularly doppler navigation antennas. It is interesting to note in addition that the effect of change in ϵ_r and length is in the opposite sense and tend to reduce the overall effect. There is significant permanent change in θ due to stress relief although as noted previously the material can be stabilised before etching. It is expected that sidelobe level changes will also occur and that these will be greater than in the example given above. There will also be some cold flow of the dielectric, particularly when under pressure for example near a connector or bolt, but there is no available data on the magnitude involved.

Etching errors are significant in arrays, particularly if they are of series fed travelling wave form. Hall and James (1978) have measured the dimensional errors

Table 8.6   *Typical manufacturing tolerances in combline production*

Production stage	Error mm	Average percentage error
x 3 full size computer drawn master	± 0·03	± 1
full size film mask	± 0·02 (random) +0·005 (correlated)	± 2 +½ (error in reduction ratio)
etching	± 0·04 (random) −0·08 (correlated)	± 4 −8 (undercutting)

in mask production and etching of comb line arrays. In the process a 3 × full size master is computer drawn and then photographically reduced to produce a full size mask. Table 8.6 gives the errors obtained together with those in the final array. It can be seen that significant errors occur in the computer drawing and photo reduction. Smaller errors will be obtained with the direct computer controlled mask production methods now available. However, larger errors are obtained upon etching, with a substantial but repeatable undercut. The effect of these errors on comb line performance has been simulated by calculating the radiation pattern of the comb with finger widths as measured on the actual antenna. At 17 GHz, the −28 dB design sidelobe level rose to −24·6 dB. About 1·5 dB of this rise can be attributed to the finite pen size used in the graphic production, as the design calls for fingers considerably narrower than can be produced. The narrowest lines repeatedly produced on comb line production are 0·15 mm wide.

8.2 Substrate surface-wave generation

In addition to tolerance effects the scattering of substrate surface waves in microstrip antennas will product further performance degradations. No satisfactory way of measuring such waves has yet been reported. However, theoretical estimates have been given in Chapters 3 and 9 and this allows the possible radiation pattern perturbations to be assessed. The ratio of surface-wave generation to radiation for an open-circuit termination, S_L, is suggested in Chapter 3 to form a pessimistic limit to the lowest sidelobe obtainable in an array of such sources. This assumes that the scattered substrate surface waves all add up in phase in some direction away from the main beam. The measured results for arrays on $h = 0·793$ mm, $\epsilon_r = 2·32$ substrates show that sidelobes can be obtained well below the $S_L = -15$ dB level, given in Fig. 3.16(b) for this substrate. This indicates that there is some degree of incoherence in the scattered waves. An optimistic limit of sidelobe perturbation is given by Elliott (1959) where the maximum mean rise in sidelobe level, ΔS, due to random perturbations in a Dolph-Tchebycheff distribution is

$$\Delta S = 10 \log_{10} [1 + 10^{D/10} \cdot 2\sigma^2 \cdot F] \tag{8.6}$$

where D is the design sidelobe level in dB, σ is the standard deviation of the perturbation and F is a function of the aperture distribution. For the array quoted above with $D = -30\,dB$, $\Delta S = 0.23\,dB$. $D -- \Delta S = -29.77\,dB$ is thus an indication of the optimistic sidelobe level for this case while $S_L = -15\,dB$ forms the most pessimistic; measured results indicate that the actual level is somewhere below $-20\,dB$.

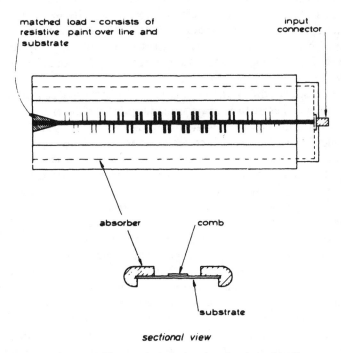

Fig. 8.1   *Comb line for h = 1·59 mm substrate showing absorbent shielding*

Experimental evidence of surface-wave scattering is demonstrated in an array similar to that described above but produced on a $h = 1.59\,mm$ thick substrate, as shown in Fig. 8.1. Sidelobes of approximately $-17\,dB$ were initially obtained; these were reduced to below $-20\,dB$ by shielding the edges of the substrate with absorbent material. In contrast, placing absorber on the edges of an array made on the $h = 0.793\,mm$ substrate had only small effect on the radiation pattern. These qualitative results indicate an increased surface wave generation for thicker substrates and together with measured radiation conductances for these substrates, Fig. 5.34, substantiate that there is an optimum substrate thickness for maximum pattern control and radiation efficiency. This then confirms, as discussed in Chapter 3, that operational 'windows' exist in plots of substrate thickness or frequency parameter below which inefficient operation is obtained and above which pattern degradation due to substrate surface wave scattering is excessive. No way of measuring the ratio of surface wave to radiated power has yet been reported.

8.3 Summary comments

In this chapter we have seen how tolerance and surface-wave effects prevent optimal performance being realised in practice; however the degree of degradation cannot be calculated to fine enough limits at present. For patch antennas, the tolerance effect on the resonant frequency is important; indeed, it was shown that for thin substrates at low frequencies tolerance effects mean that the resonant frequency cannot be guaranteed to vary in manufacture by less than its own bandwidth. The system implications here are serious and to avoid the problem, continuous monitoring of material and processing parameters may be necessary. In addition etching tolerances, shrinkage due to stress relief and cold flow also need to be taken into account.

In arrays, a further critical parameter is aperture distribution control. The effect of tolerances here is harder to estimate as the substrate parameter error distribution across the board is unknown. Calculations do suggest that this will be a significant source of sidelobe level increase. Beam pointing errors may also occur that will be important in, for instance, doppler navigation antennas. Errors here mean that each antenna's radiation pattern must be measured before use, implying increased unit costs. Etching tolerances are particularly important in arrays where thick and thin lines are used such as the comb line; discontinuity arrays may be better here.

The lower limit to which sidelobes can be reduced in the presence of substrate surface waves is not precisely known. However, the existence of operational windows, as predicted in Chapter 3, bordered on one hand by increased surface wave generation and on the other by reduced efficiency is experimentally confirmed.

As regards tolerances it is concluded that for microstrip antennas, such problems are worse than for many other types of antenna. Tolerance sensitivity may well be traded for efficiency in broad beam types but where good pattern control is required it remains a fundamental limit. Tolerance control may in future be improved but this is expected to be at the expense of simple and cheap production methods and low priced substrates.

8.4 References

ELLIOTT, R. S. (1959): 'Mechanical and electrical tolerances for two-dimensional scanning arrays', *IRE Trans.*, **AP-6**, pp. 114–120

GOEDBLOED, W., HIEBER, H., and VAN NIE, A. G. (1978): 'Ageing tests on microwave integrated circuits', *Rad. & Electron. Eng.*, **48**, 1/2, pp. 13–22

HALL, P. S., and JAMES, J. R. (1978): 'Microstrip antenna arrays'. Final Report on Research Agreement AT/2160/033RL, Royal Military College of Science, Swindon

NORWICKI, T. E. (1979): 'Microwave substrates present and future'. Proc. workshop on printed antenna technology, New Mexico, USA, pp. 26-1 to 26-12

PETRIE, E. M., and GROVE, R. (1971): 'Dimensional stability of stripline materials'. Proc. 10th Electrical Insulation Conference, Chicago, pp. 179–183 (IEEE, New York)

TRAUT, G. R. (1979): 'Clad laminates of PTFE composites for microwave antennas'. Proc. workshop on printed antenna technology, New Mexico, USA, pp. 27-1 to 27-17

Chapter 9

Recent advances in microstrip antenna analysis

The analyses of microstrip antenna element configurations discussed in earlier Chapters were based on some degree of approximation concerning the current and field distributions. The results obtained show reasonable agreement with practical results in some cases, but various adjustments have to be made to the parameters used in the calculations to improve agreement. Examples of this are the use of effective widths of open circuit transmission lines rather than the physical width to calculate the radiation conductance (Section 3.1), and the allowances for fringe fields in calculating the resonant frequency of patch antennas (Section 4.5.2). In this Chapter, recent advances in the mathematical analysis of microstrip antennas are outlined and the extent to which they model the precise physical action is particularly discussed together with likely sources of error.

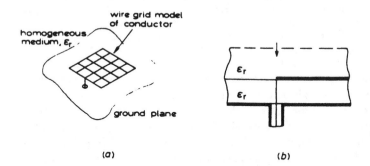

(a)   (b)

Fig. 9.1 *Homogeneous dielectric medium (ϵ_r) wire-grid model of microstrip patch antenna (Agrawal and Bailey, 1977)*
  a Model configuration
  b Empirical compensation for practical substrate

The variational analysis of the open-circuit line discussed in Section 3.4 gives an improved theory for this elemental radiator which takes account of the presence of the substrate. If the substrate were not present the various embodiments of moment method techniques (Harrington, 1968) could be directly applied to complete antenna structures. This fact has been used by Agrawal and Bailey (1977) and by Newmann and Tulyathan (1979) in microstrip patch analyses which may be regarded as intermediate between the engineering analyses and the desired exact analysis since the presence of the dielectric is in each case only approximately accounted for. Agrawal and Bailey approach the analysis by using a standard wire-grid modelling computer programme to calculate the properties of a patch antenna in a homogeneous dielectric medium of relative permittivity ϵ_r (Fig. 9.1(a)). The results are then adjusted for the true case by factors obtained by empirical studies of the variation in the characteristics of a microstrip antenna with overlaying dielectric of various thicknesses (Fig. 9.1(b)). In contrast, Newmann and Tulyathan carry out an analysis of a rectangular patch antenna with a probe feed when in a homogeneous air medium using three current basis functions whose directions are shown in Fig. 9.2(a). The dielectric is then accounted for by referring the calculated radiation impedance to the edge of the patch and using a microstrip transmission line model (Derneryd, 1977) to predict the final input impedance characteristics (Fig. 9.2(b)). However, neither of these analyses can be regarded as being fully satisfactory owing to the approximate methods of incorporating the effects of the dielectric substrate.

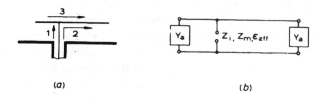

(a)   (b)

Fig. 9.2   *Homogeneous dielectric medium (ϵ_0) current mode model of microstrip patch antenna (Newmann and Tulyathan, 1979)*
  a Model configuration showing current mode positions
  b Network method of compensation for practical substrate

The available literature on more rigorous analyses of microstrip antennas is limited at present to work considering compact antennas whose dimensions are less than a wavelength, and which may be essentially regarded as resonant elements. A brief outline of the work, which can be conveniently split into four separate categories, is given in the following Sections of this Chapter, followed by a comparison of their characteristics and discussion of possible extensions. Table 9.1 summarises the structures analysed, methods used and results available for each category of analysis. The derivation of each of these analyses requires the establishment of relationships between current sources and field distributions in the presence of the substrate and ground plane. This requires the boundary condition relationships between field components at the substrate surface to be applied, which is most easily implemented by means of transforms to spectral domains. Although the details of such derivations will vary depending on the structure considered, many of the functions generated are common to all. The notations used in the description of the analyses given in this Chapter therefore vary from those in the source references in order to clarify the relationships between them.

Table 9.1 *Summary of more rigorous analyses of microstrip antennas*

Section	Structure	Analysis method	Results available in referenced literature
9.1	Wire dipole at substrate surface	Segmented wire with central source, single basis function on each.	Dipole impedance, mutual impedance.
9.2	Narrow rectangular strip	TEM-mode transmission-line current on each half of centre-fed strip.	Impedance, polar diagram, surface-wave/free-space power ratio.
9.3	Circular patch	Orthogonal mode current expansion, one assumed as distributed source.	Resonant frequency, bandwidth, polar diagrams, current distribution, surface-wave/free-space power ratio.
9.4	Rectangular patch	Basis current mode expansion, incorporating edge singularity in modes.	Radiation pattern from single mode.

Note: More details on the above and other similar analyses can be found in a special issue of the IEEE Transactions, AP-29, No. 1, 1981 devoted to microstrip antennas.

9.1 Wire antenna analysis

Early work on the calculation of fields due to sources radiating in the presence of a
ground plane with a dielectric slab was performed from the point of view of excitation of surface-wave modes. Tsandoulas (1969), for example, calculates the
radiation field of a half-wave dipole antenna parallel to the surface of a grounded
dielectric slab (Fig. 9.3) assuming a sinusoidal current distribution on the antenna.
Although some comparative results were presented for free-space and surface
wave power radiation, no attempt was made to calculate the input impedance of

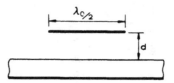

Fig. 9.3 $\lambda_0/2$ *dipole over microstrip substrate*

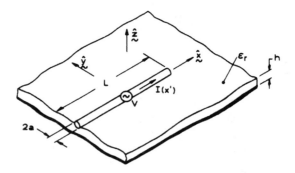

Fig. 9.4   *Wire antenna at microstrip substrate surface (Rana and Alexopoulos 1979a, b)*

the antenna. More recently, Rana and Alexopoulos (1979a, 1979b) have given a
moment method analysis of both a single dipole and a parallel pair of dipoles
(Fig. 9.4) with data on impedances being given. This analysis uses a Hertzian vector
formulation to derive the Green's function relating the electric field at a point on
the surface of the substrate to an exciting Hertzian dipole. The axial electric field
due to the total antenna current is then given by Pocklington's equation:

$$E_x(x,y,z) = \int_L I(x')\left[k_0^2 \pi_x + \frac{\partial^2 \pi_x}{\partial x^2} + \frac{\partial^2 \pi_z}{\partial x \partial z} \right] dx' \qquad (9.1)$$

where π_x and π_z are components of the Hertzian vector due to the Hertzian dipole
and are given by:

$$\pi_x = 2u \int_0^\infty J_0(\lambda\rho) \exp\left(-\eta\{z-h\}\right) \frac{\lambda d\lambda}{D_e(\lambda)} \qquad (9.2a)$$

$$\pi_z \; = \; -2(\epsilon_r - 1)u \int_0^\infty J_1(\lambda \rho) \cos{(\phi)} \exp{(-\eta\{z - h\})} \frac{\lambda^2 d\lambda}{D_e(\lambda)D_m(\lambda)}$$

$$(9.2b)$$

with

$$\rho \; = \; \sqrt{(x - x')^2 + y^2}$$

$$D_e(\lambda) \; = \; \eta + \eta_d \coth{(\eta_d h)}$$

$$D_m(\lambda) \; = \; \epsilon_r \eta + \eta_d \tanh{(\eta_d h)}$$

$$u \; = \; 1/(4\pi j \omega \epsilon_0), \qquad \eta \; = \; \sqrt{\lambda^2 - k_0^2}$$

$$\cos{(\phi)} \; = \; (x - x')/\rho, \qquad \eta_d \; = \; \sqrt{\lambda^2 - \epsilon_r k_0^2}$$

In the application of the moment method solution the antenna is divided into N equal segments, with the current magnitude on each segment being assumed to be independent and of the form:

$$I(x') \; = \; \sum_{i=1}^{N} a_i I_i(x')$$

$$I_i(x') \; = \; \frac{\sin{(k_0\{d - |x' - x_i|\})}}{\sin{(k_0 d)}}; \qquad |x' - x_i| \leqslant d \qquad (9.3a)$$

$$= \; 0; \qquad\qquad\qquad |x' - x_i| > d \qquad (9.3b)$$

where $d = L/N$ and x_i is the centre of the ith segment.

The antenna current distribution and input impedance can then be derived by applying current continuity conditions between segments and the condition that:

$$\int_{x_i - d/2}^{x_i + d/2} E(x') F(x') dx' \; = \; 0 \qquad (9.4)$$

on the antenna for each segment. The testing functions $F(x')$ are chosen in accordance with the Galerkin procedure to be identical in form to the current distributions as given by eqn. 9.3. The evaluation of eqn. 9.4 requires the wire to be of finite size to avoid an infinite reactance resulting, and Rana and Alexopoulos assume a round wire in their papers. Typical results obtained by them for current distributions and impedance of isolated dipoles are shown in Fig. 9.5, and mutual impedance data for broadside and endfire coupling in Fig. 9.6. The general impedance behaviour has characteristics similar to those of dipoles in free space although the components are somewhat smaller as would be expected from the presence of the groundplane. The mutual impedance data had asymptotic behaviour which indicated that surface wave coupling was generally weaker, and this supports the conclusions drawn in Section 5.3.2 concerning mutual coupling in arrays. No polar diagram information was presented by Rana and Alexopoulos, although this would of course be readily derivable from the current distribution. An application of this analysis method to a Yagi-Uda array on a microstrip substrate has also been discussed by Alexopoulos *et al.* (1980), with some results on front to back ratio and driven element impedance being given.

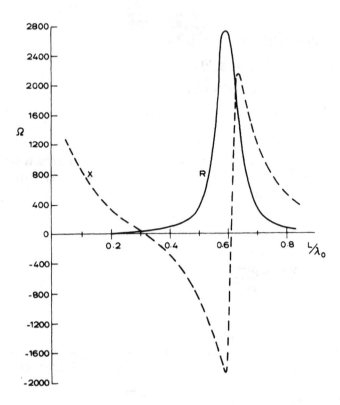

Fig. 9.5   *Input impedance of centre-fed wire antenna on microstrip substrate (After Rana and Alexopoulos, 1979b)*
———— resistance
— — — reactance
Substrate height $h = 0.1016\lambda_0$, permittivity $\epsilon_r = 3.25$
Dipole diameter $2a = \lambda_0/10\,000$

In a real situation, a microstrip antenna will have a flat aspect. There are well known relations governing equivalences between wires of various cross-sections (King, 1956), in particular a flat strip of width w has an equivalent radius r given by

$$r = w/4 \tag{9.5}$$

Whereas in the problem of wire antennas in free space the use of this equivalence is normally only restricted by the wire dimensions being small compared to the wavelength, in the microstrip case the close presence of the ground plane means the analysis could only be applied to strips which were also narrow compared to the substrate thickness. Since this is not usually the case in microstrip antenna designs, the analysis as it stands is of restricted usefulness. It may be possible in the future to extend this method to antennas whose transverse dimensions are large compared to the substrate thickness by a wire grid equivalence to the antenna, similar to that discussed earlier in this Chapter. However, the difficulties involved in satisfactory wire-grid modelling of large structures in free space are great (Ramsdale, 1978) and the present problem is further complicated by the involved Green's functions occuring in the integral eqn. 9.1.

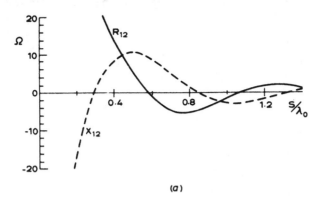

(a)

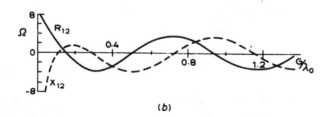

(b)

Fig. 9.6 *Mutual impedance of centre-fed wire antennas on microstrip substrate (After Rana and Alexopoulos, 1979b)*
———— resistance
— — — reactance
a Broadside coupling
  S = separation between centres of antennas, $L = 0.333\lambda_0$
  Substrate $h = 0.1016\lambda_0$, $\epsilon_r = 3.25$
b Colinear coupling
  G = gap between ends of antennas, $L = 0.25\lambda_0$
  Substrate $h = 0.1016\lambda_0$, $\epsilon_r = 3.25$

9.2 Narrow rectangular strips

Uzunoglu *et al.* (1979) have given an analysis of a narrow strip microstrip antenna with a central driving source, which starts by formulating the dyadic Green's function of a Hertzian dipole by a plane wave spectrum representation (Fig. 9.7):

$$G(\mathbf{r}/\mathbf{r}') = \frac{30j}{\pi k_0} \int \int d\lambda_x d\lambda_y \frac{\exp(j\boldsymbol{\lambda} \cdot \{\mathbf{r} - \mathbf{r}'\})}{D_e(\lambda)D_m(\lambda)} \mathcal{M} \tag{9.6}$$

where $\mathcal{M}$ is the matrix defined by

$$\mathcal{M} =$$

$$\begin{pmatrix} (k_0^2 - \lambda_x^2)\eta_d \tanh(\eta_d h) + (\epsilon_r k_0^2 - \lambda_x^2)\eta & -\lambda_x \lambda_y (\eta + \eta_d \tanh(\eta_d h)) \\ -\lambda_x \lambda_y (\eta + \eta_d \tanh(\eta_d h)) & (k_0^2 - \lambda_y^2)\eta_d \tanh(\eta_d h) + (\epsilon_r k_0^2 - \lambda_y^2)\eta \end{pmatrix}$$

and

$$\boldsymbol{\lambda} = \hat{x}\lambda_x + \hat{y}\lambda_y \qquad \begin{aligned} \mathbf{r} &= \hat{x}x + \hat{y}y \\ \mathbf{r}' &= \hat{x}x' + \hat{y}y' \end{aligned}$$

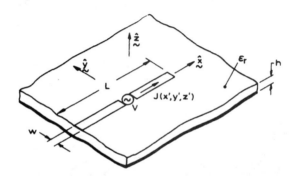

Fig. 9.7   *Narrow rectangular strip on microstrip substrate surface (Uzunoglu et al., 1979)*

This formula is then used in a variational expression for the input impedance of the antenna in conjunction with the assumed current distribution $\mathbf{J}$ (Collin, 1961)

$$Z_{in} = \frac{\int_A \int_A \mathbf{J}(\mathbf{r}) \cdot (G(\mathbf{r}/\mathbf{r}')\mathbf{J}(\mathbf{r}')) \, d\mathbf{r} \, d\mathbf{r}'}{I_0^2} \tag{9.7}$$

where I_0 is the total input current and A is the area of the antenna conductor. The advantage of the formulation of eqn. 9.6 is that it may relatively easily be applied to the case of a flat strip, and Uzunoglu *et al.* give the expression derived from eqn. 9.7 for a strip dipole of width w carrying a single assumed current distribution

$$\begin{aligned} \mathbf{J}(\mathbf{r}) &= \hat{x} \, |y|I_0 \sin\left(\beta\left\{\frac{L}{2} - |x|\right\}\right); \qquad |x| \leqslant \frac{L}{2}, \quad |y| \leqslant w/2 \\ &= 0; \qquad\qquad\qquad\qquad\qquad\quad \text{elsewhere} \end{aligned} \tag{9.8}$$

where $\beta = 2\pi\sqrt{\epsilon_e}/\lambda_0$ is the effective propagation constant of a microstrip transmission line of width w on the same substrate.

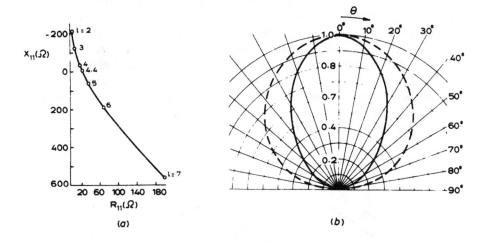

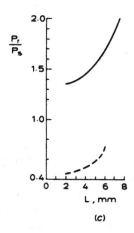

Fig. 9.8 *Characteristics of centre-fed rectangular strip antenna on microstrip substrate (After Uzunoglu et al., 1979)*

    a Input impedance
       Substrate $h = 3.048$ mm, $\epsilon_r = 3.25$, $w = 0.5$ mm
       Frequency $= 10$ GHz
    b Polar diagrams
       ——— $|E_\theta|$ in $\phi = 0°$ plane
       — — — $|E_\phi|$ in $\phi = 90°$ plane
       Substrate $h = 1$ mm, $\epsilon_r = 9.9$, $L = 7.5$ mm, $w = 0.5$ mm
       Frequency $= 10$ GHz
    c Ratio of free-space power to surface-wave power
       ——— Substrate $h = 3.048$ mm, $\epsilon_r = 3.25$, $w = 0.5$ mm, frequency $= 10$ GHz
       — — — Substrate $h = 2$ mm, $\epsilon_r = 9.9$, $w = 0.5$ mm, frequency $= 10$ GHz

Results of impedance calculations with this assumed current distribution are shown in Fig. 9.8(a), with similar general characteristics to those of Fig. 9.5 for the moment method wire analysis discussed in Section 9.1. Uzunoglu *et al.* also gave the results of a polar diagram calculation (Fig. 9.8(b)) but for an alumina substrate which is not commonly used for microstrip antennas and thus is of limited value. Finally, results were also given of the ratio of the free space radiation and surface wave components of the input resistance of the antenna (Fig. 9.8(c)) showing that the surface wave component was very strong for the $\epsilon_r \sim 3\cdot25$ case and dominated the resistance for the $\epsilon_r \sim 9\cdot9$ case.

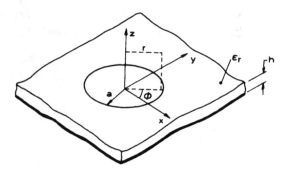

Fig. 9.9   *Circular disc microstrip patch antenna (Wood, 1981)*

9.3 Circular patch

The first study of an open circular microstrip goemetry was carried out by Pintzos and Pregla (1978), who considered the resonant frequency of the ring resonator structure when the strip width was of the same order as the substrate thickness. No consideration was given to the radiation properties of the rings so that the results given are of little interest from the point of view of antenna analysis, furthermore such elements typically have narrow bandwidth and low efficiency compared to other microstrip antenna elements. Analysis of the circular microstrip patch of Fig. 9.9 has been given by Chew and Kong (1980) and Wood (1981), the first of these simplifying the problem by considering only the resonant frequency of axisymmetric modes ($\partial/\partial\phi = 0$) which again are of little practical interest for antenna design because they give no radiation in the direction of the disc axis. The results presented by Wood are restricted to the dominant mode which is the one normally used in antenna design, however characteristics such as antenna bandwidth and radiation of surface waves are dealt with in addition to resonant frequency and the treatment given here will essentially follow that analysis. The analysis requires derivation of the relationship between the tangential electric field and a current distribution at the substrate surface, which because of the need to use a cylindrical coordinate system is complicated by the resulting occurrence of transforms related to the Hankel transform. Dual transform equations are obtained since the relationship is between two components of the electric field and two components of current, these being (Wood, 1981)

$$\int_0^\infty \left\{ jE_r(r)J_n'(\lambda r) + E_\phi(r)\frac{nJ_n(\lambda r)}{\lambda r} \right\} rdr =$$

$$\frac{\eta\eta_d \tanh(\eta_d h)}{\omega\epsilon_0 D_m(\lambda)} \int_0^\infty \left\{ jI_\phi(r)\frac{nJ_n(\lambda r)}{\lambda r} - I_r(r)J_n'(\lambda r) \right\} rdr \qquad (9.9a)$$

$$\int_0^\infty \left\{ jE_r(r)\frac{nJ_n(\lambda r)}{\lambda r} + E_\phi(r)J_n'(\lambda r) \right\} rdr =$$

$$-\frac{\omega\mu_0}{D_e(\lambda)} \int_0^\infty \left\{ jI_\phi(r)J_n'(\lambda r) - I_r(r)\frac{nJ_n(\lambda r)}{\lambda r} \right\} rdr \qquad (9.9b)$$

where $J_n(x)$ is the Bessel function of the first kind of order n.

The method used is to assume that the current distribition on the disc may be represented by the orthogonal mode series

$$\mathbf{I} = e^{jn\phi} \sum_{p=1}^P a_p \left\{ \hat{\mathbf{r}}\frac{nJ_n(\gamma_p r)}{\gamma_p r} + \hat{\phi}jJ_n'(\gamma_p r) \right\}$$

$$+ e^{jn\phi} \sum_{q=1}^Q a_q \left\{ \hat{\mathbf{r}}J_n'(\eta_q r) + \hat{\phi}j\frac{nJ_n(\eta_q r)}{\eta_q r} \right\}; \qquad r \leqslant a \qquad (9.10a)$$

$$\mathbf{I} = 0; \qquad\qquad\qquad\qquad\qquad\qquad r > a \qquad (9.10b)$$

The η mode which corresponds to the current distribution obtained from the magnetic wall model analysis for the mode of interest is assumed to be the driving current which supplies the power radiated by the antenna, and the remaining modes are then assumed to be free modes which allow the true current distribution to be better approximated. A matrix equation may be set up starting from the relation

$$\int_S \mathbf{E} \cdot \mathbf{I}_s^* ds = 2P_s \qquad (9.11)$$

where $\mathbf{I}_s$ is a current source on the surface S, $\mathbf{E}$ is the electric field in the region and P_s is the complex power supplied by the current source. Thus, if the field $\mathbf{E}$ calculated by substituting the total disc current $\mathbf{I}$ of eqn. 9.10 in eqn. 9.9 is separately combined in eqn. 9.11 with each of the current modes, a set of expressions for the power (P_p, P_q) of the modes results. Using the assumed conditions that $(P_p, P_q) = 0$ except for the driving source the following matrix equation is obtained:

$$k_{11}a_1a_1^* + k_{12}a_2a_1^* + \ \dots \ k_{1M}a_Ma_1^* \ = \ 0$$
$$k_{21}a_1a_2^* + k_{22}a_2a_2^* + \ \dots \ k_{2M}a_Ma_2^* \ = \ 0$$
$$\vdots \qquad\qquad \vdots \qquad\qquad\qquad \vdots \qquad\qquad \vdots$$
$$k_{M1}a_1a_M^* + k_{M2}a_2a_M^* + \dots \ k_{MM}a_Ma_M^* \ = \ 2P_M \qquad (9.12)$$

where for convenience the two indices p, q have been combined in a single series m, with the index $M = P + Q$ being assigned to the driving source. The matrix elements are given by:

$$k_{m'm} = -2\pi j \int_0^\infty \left[\frac{\eta_d \tanh(\eta_d h)}{\omega \epsilon_0 D_m(\lambda)} \times \right.$$

$$\left\{ \int_0^\infty \left(j I_{\phi m}(r) \frac{n J_n(\lambda r)}{\lambda r} - I_{rm}(r) J_n'(\lambda r) \right) r dr \right\} \times$$

$$\left\{ \int_0^\infty \left(j I_{\phi m'}^*(r) \frac{n J_n(\lambda r)}{\lambda r} + I_{rm'}^*(r) J_n'(\lambda r) \right) r dr \right\} - \frac{\omega \mu_0}{D_e(\lambda)} \times$$

$$\left\{ \int_0^\infty \left(j I_{\phi m}(r) J_n'(\lambda r) - I_{rm}(r) \frac{n J_n(\lambda r)}{\lambda r} \right) r dr \right\} \times$$

$$\left. \left\{ \int_0^\infty \left(j I_{\phi m'}^*(r) J_n'(\lambda r) + I_{rm'}^*(r) \frac{n J_n(\lambda r)}{\lambda r} \right) r dr \right\} \right] \lambda d\lambda \qquad (9.13)$$

Applying the Gauss elimination method to eqn. 9.12 gives

$$k_{11}' a_1 a_1^* + k_{12}' a_2 a_1^* + \ldots k_{1M}' a_M a_1^* = 0$$

$$k_{22}' a_2 a_2^* + \ldots k_{2M}' a_M a_2^* = 0$$

$$\vdots \qquad \vdots$$

$$k_{MM}' a_M a_M^* = 2P_M \qquad (9.14)$$

Since the a_m are current amplitudes and P_M the complex power associated with the driving mode, the final equation allows the identity:

$$Z_{in} = k_{MM}' \qquad (9.15)$$

to be made. The remaining $(M-1)$ equations may be used to calculate the amplitudes of the other modes normalised to a_M.

The input impedance of the driving mode may be used to determine the resonant frequency of the mode by varying the frequency of the driving mode until $Im(Z_{in}) = 0$. The Q factor may also be calculated by finding the frequencies f^+ and f^- for which the reactance becomes equal in magnitude to the resistance at resonance and using the standard relation

$$Q = f_r/(f^+ - f^-) \qquad (9.16)$$

Wood considered the rate of convergence of the analysis as a function of the number of modes used in the analysis and found that this was very slow, particularly for large a/h ratios. The source of this problem lay in the large number of modes required to represent the edge singularity of the current distribution on the disc,

Fig. 9.10 showing some plots of the computed results. However, the residual error of the computation due to the finite number of modes used was estimated as 1% for 40 modes which is sufficient for most practical requirements when material tolerances effects are considered (see Section 8.1). Results obtained from the numerical analysis for the resonant frequency are shown in Fig. 9.11 compared

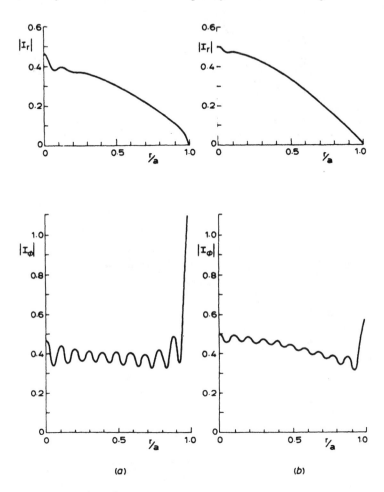

Fig. 9.10   *Current distribution on microstrip circular disc (After Wood, 1981)*
Dominant resonance ($n = 1$), 40 current modes used, $\epsilon_r = 2\cdot32$
a $a/h = 3\cdot162$
b $a/h = 31\cdot62$

with experimental results and the approximate theoretical figures given by the formula due to Shen *et al.* (1977) as given in eqn. 4.64. These show that the simple formula gives good agreement for large a/h ratios, and in view of the difficulties encountered with the convergence of the numerical analysis in this region

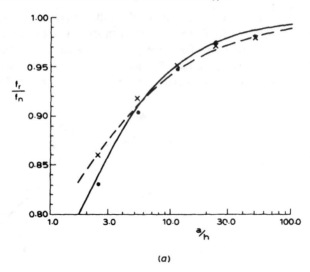

(a)

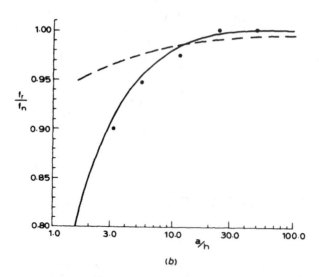

(b)

Fig. 9.11   *Resonant frequency of microstrip circular disc (After Wood, 1981)*
Dominant resonance ($n = 1$), f_n = resonant frequency using magnetic wall model, eqn. 4.29
a $\epsilon_r = 2\cdot32$
b $\epsilon_r = 9\cdot8$
————— numerical theory, 40 current modes used
— — — approximate theory, eqn. 4.64
Experimental results: x coax feed, ● microstrip feed

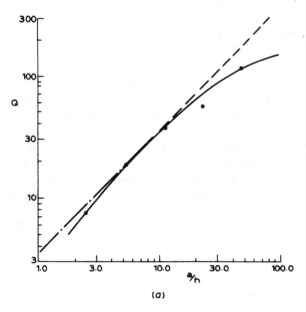

(a)

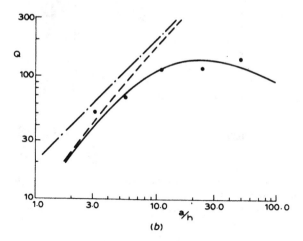

(b)

Fig. 9.12   *Q-factor of microstrip circular disc (After Wood, 1981)*
Dominant resonance (*n* = 1)
a $\epsilon_r = 2 \cdot 32$
b $\epsilon_r = 9 \cdot 8$
— — — numerical theory, 4 current modes used
———— numerical theory with allowance for conduction losses, eqn. 4.19
— · — · approximate theory, Section 4.3.2
Experimental results: • microstrip feed

there seems to be little practical advantage gained by its use. There is a useful improvement obtained at the lower a/h values, and for the higher relative permittivity.

Results of the bandwidth calculated from eqn. 9.16 are shown in Fig. 9.12, compared with experimental results and the approximate theoretical values calculated by the 'magnetic' current source·method discussed in Section 4.3.2. Agreement with experiment is good over the full range of a/h values provided the effect of conduction losses is included. The simple theory deviates most for small a/h values, and this is almost entirely due to the lack of an allowance for surface-wave radiation. The magnitude of the discrepancy suggests that the surface-waves radiation may form a significant part of the total power radiated by the antenna, and this is confirmed in Fig. 9.13 which shows the ratio of surface-wave power to total power of the antenna. As has been discussed elsewhere in this book, surface-wave radiation

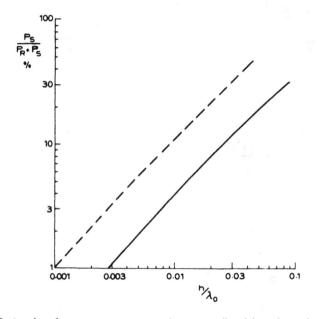

Fig. 9.13   *Ratio of surface-wave power to total power radiated by microstrip circular disc (After Wood, 1981)*
    Dominant resonance ($n = 1$), 4 current modes used in numerical theory
    ——— $\epsilon_r = 2 \cdot 32$
    — — — $\epsilon_r = 9 \cdot 8$

may form a limiting factor upon the maximum usable frequency of any microstrip substrate since the power coupled into such waves will be scattered from the substrate edges leading to degradation of polar diagrams, and it also represents a power loss in the antenna. If it is assumed that the tolerable fraction of power coupled to the surface wave is 25%, then this analysis gives upper limits to the frequency for the substrates obtained from Fig. 9.13 of $21/h$ GHz for $\epsilon_r = 2 \cdot 32$ and $6 \cdot 9/h$ GHz for $\epsilon_r = 9 \cdot 8$, where h is in millimetres. It should be noted that the proportion of the radiated power which coupled to the surface wave predicted by this numerical analysis is larger than that obtained from the variational analysis of the open circuit microstrip line discussed in Sections 3.5 and 3.6, particularly for the higher permittivity substrate. However, the results of Uzunoglu *et al.* (1979) for the surface-wave to free-space ratio of a narrow strip radiator (Fig. 9.8(*c*)) are consistent with extrapolation of the curves given in Fig. 9.13, which indicates that this is not simply due to the different geometries involved, and the discrepancy has not as yet been finally resolved.

9.4 Rectangular patch

An application of the full wave analysis to rectangular patch antennas has been outlined by Itoh and Menzel (1979) (Fig. 9.14) in which the mathematical manipulations are somewhat simpler than those of Section 9.3 since they involve Fourier

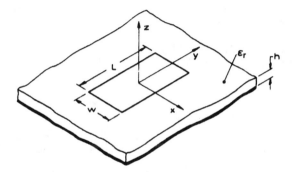

Fig. 9.14 *Rectangular microstrip patch antenna (Itoh and Menzel, 1979)*

transforms rather than the involved cylindrical transforms. Defining the general Fourier transform as

$$\Phi(\lambda_x, \lambda_y) = \int dx \int_{-\infty}^{\infty} dy \cdot \phi(x,y) \exp\left(j(\lambda_x x + \lambda_y y)\right) \tag{9.17}$$

the spectral domain relationship between the surface electric field and current distributions is given by

$$\begin{pmatrix} E_x(\lambda_x, \lambda_y) \\ E_y(\lambda_x, \lambda_y) \end{pmatrix} = \frac{j}{\omega\epsilon_0 D_e(\lambda) D_m(\lambda)} \mathscr{M}(\lambda_x, \lambda_y) \begin{pmatrix} J_x(\lambda_x, \lambda_y) \\ J_y(\lambda_x, \lambda_y) \end{pmatrix} \tag{9.18}$$

where $\mathscr{M}(\lambda_x, \lambda_y)$ is as defined at eqn. 9.6. This may be used to generate a matrix equation by noting that if $J_x'(x,y)$ and $J_y'(x,y)$ are current distributions on the patch

$$4\pi^2 \int_{-\infty}^{\infty} dx \int_{-\infty}^{\infty} dy \, J_x'(x, y) \, E_x(x, y) =$$

$$= \int_{-\infty}^{\infty} d\lambda_x \int_{-\infty}^{\infty} d\lambda_y \, J_x'(\lambda_x, \lambda_y) \, E_x(\lambda_x, \lambda_y)$$

$$= 0 \tag{9.19}$$

by Parseval's relation and a similar relationship holds between E_y and J_y'. The identity with zero is obtained by noting that tangential $\mathbf{E} = 0$ over the patch and $\mathbf{J}' = 0$ elsewhere. Thus, if a modal representation of J_x and J_y is used and eqn. 9.19 is applied for each individual current mode the matrix equation is derived. So if

$$J_x(\lambda_x, \lambda_y) = \sum_{m=1}^{M} c_m J_{xm}(\lambda_x, \lambda_y) \tag{9.20a}$$

and

$$J_y(\lambda_x, \lambda_y) = \sum_{n=1}^{N} d_n J_{yn}(\lambda_x, \lambda_y) \tag{9.20b}$$

the matrix equation derived is

$$\sum_{m=1}^{M} K_{pm}^{xx} c_m + \sum_{n=1}^{N} K_{pn}^{xy} d_n = 0; \qquad p = 1 \ldots M \tag{9.21a}$$

$$\sum_{m=1}^{M} K_{qm}^{yx} c_m + \sum_{n=1}^{N} K_{qn}^{yy} d_n = 0; \qquad q = 1 \ldots N \tag{9.21b}$$

where the elements of matrix K are given by double integrals, for example

$$K_{pn}^{xy} = \frac{j}{\omega \epsilon_0} \int_{-\infty}^{\infty} d\lambda_x \int_{-\infty}^{\infty} d\lambda_y \left[\frac{J_{xp}(\lambda_x, \lambda_y) \, \mathcal{M}_{xy}(\lambda_x, \lambda_y) J_{yn}(\lambda_x, \lambda_y)}{D_e(\lambda) D_m(\lambda)} \right]$$

In principle the resonant frequency is then established from the condition that det $|K| = 0$ since any other condition gives all the c_m and d_n as zero. In practice, the fact that the patch radiates means that this is never obtained for a real frequency, so that the calculation must either compute the complex resonant frequency and assume that the real component is the required value, or simply find the real frequency for which det $|K|$ is a minimum. An expression in simple cos or sin functions of x and y is possible, but as with the circular patch analysis of Section 9.3 a large number of modes would be required to represent the edge singularities. Itoh and Menzel suggest that modes of the form

$$J_{xm}(x, y) = \frac{\sin\left(\{r - 1\} \dfrac{2\pi x}{w}\right)}{\sqrt{\left|\dfrac{w}{2}\right|^2 - x^2}} \cdot \frac{\sin\left(\{2s - 1\} \dfrac{\pi y}{L}\right)}{\sqrt{\left|\dfrac{L}{2}\right|^2 - y^2}} \tag{9.22a}$$

$$J_{yn}(x,y) \doteq \frac{\cos\left(\{r-1\}\frac{2\pi x}{w}\right)}{\sqrt{\left(\frac{w}{2}\right)^2 - x^2}} \cdot \frac{\cos\left(\{2s-1\}\frac{\pi y}{L}\right)}{\sqrt{\left(\frac{L}{2}\right)^2 - y^2}} \qquad (9.22b)$$

will alleviate this problem since these modes incorporate a singularity. The transforms for this modal choice are combinations of Bessel functions of zero order, however, which will lengthen the computation for each mode. Furthermore, Wood (1981) found for the circular patch that the edge singularity was very localised to the edge whereas the denominators in eqn. 9.22 would lead to a comparatively slow fall away from the patch edge. Thus it is not as yet clear what the best choice of modes for particular problems is likely to be. Results by Itoh and Menzel for a single radiation pattern are shown in Fig. 9.15.

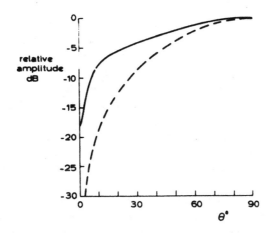

Fig. 9.15 *Radiation patterns of rectangular microstrip patch (After Itoh and Menzel, 1979)*
——— $|E_\theta|$ in $\phi = 0°$ plane
— — — $|E_\phi|$ in $\phi = 90°$ plane
Substrate $h = 1·58$ mm, $\epsilon_r = 2·32$, $w = 15$ mm, $L = 10$ mm
Resonant frequency $= 8·75$ GHz

9.5 Discussion of analytical methods

Comparison of the methods of analysing microstrip antennas in the previous two Sections tends to highlight their similarities rather than differences, as shown by the occurence of $\mathcal{M}$ in both the strip analysis at eqn. 9.6 and the rectangular patch analysis at eqn. 9.18. This is of course only to be expected since, irrespective of the detailed approach, calculation of the electric field distribution over the surface of the substrate due to a surface current source is required and this will differ only in terms of the formulation used (i.e. the Hertzian vector representation of Rana and

Alexopoulos (1979b) or the plane wave spectrum Green's function of Uzunoglu *et al.* (1979). A comparative assessment is therefore dependent more on the suitability of the various analyses for incorporating features of the microstrip antenna than their relative numerical efficiency for particular problems. The numerical analyses above are all presently restricted in application to antennas which are compact in terms of wavelengths. Difficulties will arise when arbitrarily shaped patches are considered, although there is little practical advantage gained by using varied shapes since the general performance of compact patches whose dimensions are $< \lambda_0$ are not too dissimilar.

Major analytical difficulties arise when distributed antennas of the travelling wave type are considered, such as the comb, rampart and serpent line antennas considered in Section 5.2, where mutual interaction between all parts of the antenna are possible. A precedence to this problem is the analysis of the slotted waveguide array, which is typically approached by analysing individual elements alone to determine their interaction with the line feed; network analysis is then used to predict the overall response. This method has already been applied to comb type microstrip arrays as described in Section 5.4.1 but whether it can be further refined to include mutual coupling, surface waves etc., remains to be seen; the extension to distributed arrays such as the serpent line would appear to be difficult.

An alternative to this may be the use of a moment-method wire-grid analysis based on the method discussed in Section 9.1, which has the advantage that it is suitable for a subdomain approach to the analysis which would be useful when dealing with antennas incorporating discontinuities such as T-junctions and bends. However, as discussed in Section 9.1, each wire of the model would have to be small in diameter compared to the substrate height so that for practical antennas several parallel wires each divided into segments would be needed to represent each straight strip.

Numerical accuracy and convergence of series can be identified as problem areas which influence the extent to which a particular method represents the physical situation. Round-off error is generally the least problem. However, as was shown by Wood (1981), the modelling accuracy may be poor unless large numbers of modes are used in order to obtain good representation of the current distribution of the antenna. The same situation arises in the mathematically similar analysis of the propagation constant of microstrip transmission lines discussed in Section 2.3.2 even though the line width is typically comparable with the substrate thickness. For instance, the survey by Kuester and Chang (1979) indicated a spread of as much as 15% in results of effective permittivity due to the different numbers and forms of basis currents.

It is not until the results of several alternative analyses become available that meaningful comparisons can be extracted and some deductions can now be made about the various calculations for substrate surface waves. We have at hand the James and Henderson variational analysis of Section 3.4 for an open-circuit microstrip termination and the spectral domain analysis of Uzunoglu *et al.*, Section 9.2, and Wood, Section 9.3, which both deal with isolated finite microstrip patches. All agree reasonably well on the radiation loss but the spectral domain analysis predicts an order higher level of surface wave generation. The variational analysis is in agreement with a Wiener-Hopf analysis by Angulo and Chang for the infinite width microstrip line. One explanation is that the finite patch does launch more surface waves into the substrate and the open end analysis is not applicable; the magnitude of the difference and the fact that qualitative experiments as in Section 8.2 together with measurements of sidelobe levels Chapers 5 and 6 tend to favour the lower result for surface-wave generation suggests that none of the finite patch analyses have so far modelled the excitation source in a realistic way. In contrast, the open-end analysis of Angulo and Chang is completely specified with the distant excitation defined by the incident wave, that is provided the waves reflected from the open aperture are not rereflected from the source. Our conclusion at present is that the results of both types of analyses are probably valid but their correspondence with the physical model needs careful interpretation. Similar problems have occurred with the modelling of the excitation gap in a wire dipole where large errors in the susceptance term can be obtained. For a further discussion of error effects in the analysis of Section 3.4 the reader is referred to the source paper by James and Henderson. From a design standpoint, the above uncertainty in the theoretical results for surface waves together with the difficulty in their measurement Section 8.2 means that more precise design curves in this respect will not be available in the immediate future. In the meantime we recommend the coarse parameter S_L (Section 3.5) as a rough guide to the limitations imposed by surface waves.

9.6 Summary comments

The approximate methods of calculating microstrip antennas and arrays described in previous Chapters are quite adequate for generating initial design data on resonance conditions, impedance levels, radiation patterns and efficiency etc. Experimental trimming can be successfully used in the final design stages to bring a radiating element into precise resonance, to optimise matching, adjust the beam pointing direction etc.; arrays of elements present a much more involved situation and experimental tests can be ineffective in identifying sources of unwanted radiation and optimising the radiation patterns and the operating bandwidth. Although a more complete analysis of microstrip arrays has yet to be reported advances have been brought about with respect to the analysis of isolated elements and four methods are outlined in this Chapter. The methods provide alternative data on the launching of surface waves into the substrate around a patch antenna but the results show disagreement with previous calculations for a semi-infinite microstrip line. Likely reasons for the difference are suggested and the realistic modelling of the excitation source is seen as one particular difficulty. Other aspects of error are also discussed. No doubt more progress will ensue in the near-future and existing disparities will be explained and resolved. Although the precise calculation of surface-wave generation for an isolated patch is an important step, the overall problem of calculating the behaviour of an array of dissimilar microstrip elements remains the outstanding difficulty. Having made that statement it must be taken in perspective with the fact that no complete analysis exists for a waveguide array of dissimilar slot elements. The essential point to grasp here is that the microstrip structure is further complicated by the natural wave-trapping capabilities of the substrate and it is this aspect that demands precise analysis if array practical design methods are to be sharpened. Microstrip arrays are thus likely to present challenging problems for some time to come in both measurement and mathematical analysis.

9.7 References

AGRAWAL, P. K., and BAILEY, M. C. (1977): 'An analysis technique for microstrip antennas', *IEEE Trans.*, **AP-25**, pp. 756–759

ALEXOPOULOS, N. G., KATEHI, P. B., and RANA, I. E. (1980): 'Radiation properties of microstrip Yagi-Uda arrays'. Proceedings of International URSI symposium on electromagnetic waves, Munich, pp. 211A/1–2

CHEW, W. C., and KONG, J. A. (1980): 'Resonance of the axial symmetric modes in microstrip disk resonators', *J. Math. Phys. (USA)*, **21**, pp. 582–591

COLLIN, R. E. (1961): 'Field theory of guided waves' (McGraw-Hill, New York)

DERNERYD, A. G. (1977): 'Linearly polarised microstrip antennas', *IEE Trans.*, **AP-24**, pp. 846–851

HARRINGTON, R. F. (1968): 'Field computation by moment methods' (MacMillan & Co., New York)

ITOH, T. T., and MENZEL, W. (1979): 'A high frequency analysis method for open microstrip structures'. Proceedings of the workshop on printed circuit antenna technology, New Mexico State University, pp. 10.1–10.20

KING, R. P. W. (1956): 'The theory of linear antennas' (Harvard University Press, Massachussets)

KUESTER, E. F., and CHANG, D. C. (1979): 'An appraisal of methods for computation of the dispersion characteristics of open microstrip', *IEEE Trans.*, **MTT-27**, pp. 691–694

NEWMANN, E. H., and TULYATHAN, P. (1979): 'Microstrip analysis technique'. Proceedings of the workshop on printed circuit antenna technology, New Mexico State University, pp. 9.1–9.8

PINTZOS, S. G., and PREGLA, R. (1978): 'A simple method of computing resonant frequencies of microstrip ring resonators', *IEE Trans.*, **MTT-27**, pp. 809–813

RAMSDALE, P. A. (1978): 'Wire antennas' *in* MARTEN, L. (Ed.): 'Advances in electronics and electron physics' (Academic Press) vol. 47, pp. 123–196

RANA, I. E., and ALEXOPOULOS, N. G. (1979a): 'On the theory of printed wire antennas'. Proceedings of 9th European Microwave Conference, Brighton (Microwave Exhibitors and Publishers Ltd., pp. 687–691)

RANA, I.E., and ALEXOPOULOS, N. G. (1979b): 'Printed wire antennas'. Proceedings of the workshop on printed circuit antenna technology, New Mexico State University, pp. 30.1–30.38

SHEN, L. C., LONG, S. A., ALLERDING, M. R., and WALTON, M. D. (1977): 'Resonant frequency of a circular disc printed circuit antenna', *IEEE Trans.*, **AP-25**, pp. 595–596

TSANDOULAS, G. N. (1969): 'Excitation of a grounded slab by a horizontal dipole', *IEE Trans.*, **17**, pp. 156–161

UZUNOGLU, N. K., ALEXOPOULOS, N. G., and FIKIORIS, J. G. (1979): 'Radiation properties of microstrip dipoles', *IEEE Trans.*, **AP-27**, pp. 853–858

WOOD, C. (1981): 'Analysis of microstrip circular patch antennas', *IEE Proc.*, Pt.H, **128**, pp. 69–76

Chapter 10

Other trends and possible future developments

In chapter 1, the concept of a planar antenna was introduced and the various forms currently available were noted. The microstrip planar antenna was shown to have some particular advantages over conventional antennas, such as very thin profile, low weight and low cost, together with some possible problems such as lower efficiency, narrower bandwidth operation, sidelobe and cross-polarisation levels, tolerance control and substrate surface waves. These properties were then addressed throughout the main text which covered the underlying theory of operation, and state-of-the-art design of the basic microstrip antenna forms. In this final chapter, some variants of microstrip antennas are presented which highlight supplementary trends in microstrip antenna development, in addition to continued improvements in the performance of basic designs. These variants fall roughly into groups according to the particular property that is being exploited, perhaps at the expense of other aspects that are unimportant for the particular application in mind. Section 10.1 describes some variations on the basic resonant radiator to create dual-frequency operation. Section 10.2 discusses multilayer techniques that allow improved bandwidth or efficiency without increased feed radiation, at the cost of a more complex and thicker structure. Hybrid arrangements are then given in Section 10.3 where microstrip, both as a radiating element and a transmission medium, is combined with other antenna forms. The ability of microstrip to conform to curved body shapes is then discussed in Section 10.4 on conformal microstrip antennas. Finally, Section 10.5 takes a brief look at the future.

10.1 Variations on the basic resonant element

10.1.1 Dual-frequency elements

By attaching small tabs to the edge of a disc resonant radiator, both dual-frequency operation and some improvement to the low-angle radiation can be achieved and practical examples of this are described by McIlvenna and Kernweis, (1979). Fig. 10.1(a) shows a disc with a single tab. Here the ratio of radiated field in the $\theta = 0°$ and $\theta = 90°$ directions in the E-plane is 7 to 8 dB compared with 13 dB for a similar disc with no tab. The *H*-plane radiation pattern is not affected. Simultaneously with this pattern change the resonant frequency is reduced by about 21% for a 3 GHz model. McIlvenna also notes that such low-angle radiation improvements can be obtained by placing a plain disc in a linear array environment with, for example, a

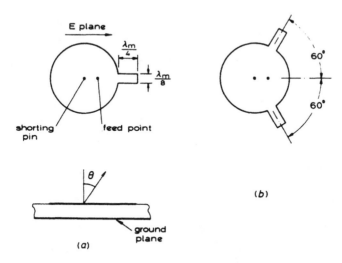

Fig. 10.1   *Microstrip disc-tab elements*
    a Single-tab type
    b Dual-tab type

$0.4\lambda_0$ spacing. It was also noticed that the addition of a single tab produced a good input VSWR at several higher frequencies in addition to the dominant mode resonance. These corresponded to higher-order modes but as expected the radiation patterns were poor. However, by using two tabs, an example being shown in Fig. 10.1(*b*), good VSWR and patterns were observed at two frequencies separated by about 50%. With more than two tabs, multiple resonances were obtained but patterns and efficiencies were poor. Tuning of the resonant frequency was suggested by electronic switching of stubs of varying length.

Patch tuning and double resonances have also been demonstrated on rectangular patches, (Kerr, 1979). Here the patch, shown in Fig. 10.2 has both an input and an output line, between 20 and 42 mm long. The output line is either short- or open-circuited. By optimising the length of the output line two resonances were observed with either termination, about 10% apart in frequency at 1·5 GHz, thus allowing dual band operation. Also by changing the length of the output line the patch can be operated at a frequency in the range from 1·05 GHz to 1·5 GHz, by using either the short- or open-circuit termination or either of the double resonances. Radiation patterns are well behaved over this bandwidth and the gain varies from 5·5 dB to 7·5 dB at the lowest and highest frequencies, respectively.

Kerr (1977) has also shown that dual-frequency operation can be obtained by placing a high-frequency resonator, for example, an *X*-band notch fed patch (Kaloi,

1976), in a cutout in a low-frequency (*L*-band) patch. This is shown in Fig. 10.3 where the two patches are orthogonally polarised. It should be noted that the feed through for the *X*-band patch is located at the voltage null at the centre of the *L*-band patch. When used as a feed for a reflector, the arrangement resulted in a

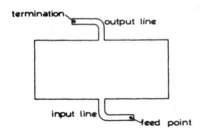

Fig. 10.2   *Terminated microstrip patch radiator*

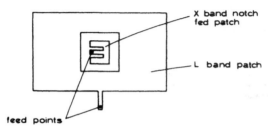

Fig. 10.3   *Dual-frequency microstrip radiator*

2 dB increase in sidelobe level compared to the −18 dB level obtained for the *X*-band patch alone.

Dual-frequency operation of an air-spaced microstrip antenna with a single feed point has been demonstrated by Sanford and Munson (1975), Fig. 10.4(*a*). The inner patch is linked to the outer one through coaxial stub coupling circuits as shown in Fig. 10.4(*b*). The stub inner is connected to the inner patch, resonant at 324 MHz. The stub outer is connected to the outer patch whose outside dimensions correspond to a patch resonant at 162 MHz. The stub length *L* is tuned to be $\lambda_c/2$ at 324 and $\lambda_c/4$ at 162 MHz, respectively, where λ_c is the wavelength in the coaxial line. It thus presents an open- and short-circuit, respectively, at point A at these frequencies. This provides a frequency selective coupling mechanism and dual-frequency operation results. The single feed is located on the diagonal at point B; circular polarisation is obtained by tuning the patch dimensions to be slightly off-square (see Chapter 7).

10.1.2 Coplanar stripline

Radiators similar to the microstrip patch have been realised in coplanar stripline backed by a ground plane (Greiser, 1976). The form of the transmission medium is shown in Fig. 10.5(*a*). The stripline is flanked by two ground planes on the upper surface forming narrow gaps along the line edge. The stripline is fed as in a normal microstrip configuration but, in addition, at the feed point, the upper ground planes are short-circuited to the lower by shorting pins. The transmission mode is

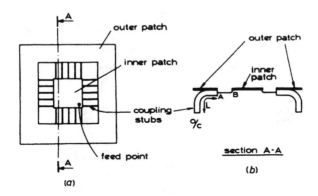

Fig. 10.4   *Dual-frequency microstrip patch radiator*
 a Patch layout
 b Coupling stub details

thus confined largely to the gaps although to a less extent than in copolar stripline with no lower ground plane; in the latter, (Wen, 1969), the centre strip is fed against the two coplanar ground planes. The form described by Greiser shown in Fig. 10.5(*b*) thus seems to be midway between true coplanar stripline and microstrip, and shows characteristics similar to those of microstrip patches. In particular, however, cross-polarisation is significantly reduced, being measured at $-20\,dB$ compared to about -10 to $-15\,dB$ for microstrip patches. Greiser attributes this to the tight binding of the wave to the slot. Low coupling between adjacent feed lines is also claimed although no measurements have been made.

10.2 Multilayer techniques

Up to this point we have considered the microstrip antenna as being constructed on a single substrate with a single etching operation forming the required pattern of conductors on the top surface with a ground plane on the lower. As we have seen, this has produced a simple structure having useful properties but essentially limited either in bandwidth or efficiency. If the single substrate requirement is relaxed then this added degree of freedom allows the trade-offs of increased height and complexity against improved design flexibility and performance to be examined. This Section, then, surveys what has so far been achieved with multilayer microstrip antenna techniques.

10.2.1 Directly coupled forms

The space occupied by two microstrip patches operating at different frequencies can be reduced by mounting the high-frequency patch on top of the low-frequency one to form the 'piggyback antenna' (Jones, 1978). As shown in Fig. 10.6, the

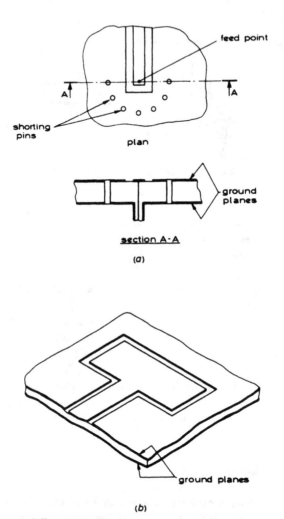

Fig. 10.5   *Coplanar stripline antenna*
   a Coplanar stripline transmission medium and feed configuration
   b Radiating element

upper conductor of the $\lambda_m/2$ microstrip patch, resonant at 0·99 GHz, forms the ground plane of a $\lambda_m/4$ short-circuited patch resonant at 1·14 GHz. The lower patch is fed in the normal way by a coaxial feed through connector, whilst the outer of the coaxial line feeding the upper patch shorts the lower patch and then connects to the upper patch conventionally. This shorting post may be placed at a voltage null of the lower patch or can be used as an inductive tuning post. The electrical separation of the two antennas is reported to be greater than $-30\,dB$.

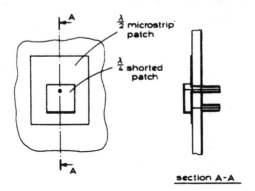

Fig. 10.6 *Microstrip piggyback antenna*

10.2.2 Electromagnetically coupled forms

If a patch radiator is placed above a microstrip line then coupling will occur due to interaction of the fringe fields. This is similar to gap coupling of coplanar patches to feed lines resulting in either impedance matching of single patches, Chapter 4, or the formation of series arrays of capacitively coupled resonators, Section 5.2.1; however, the vertical displacement now provides a significant increase in the range and controllability of the coupling. Oltman (1977, 1978) has described several forms of the overlaid electromagnetically coupled resonator. Fig. 10.7(a) shows resonators coupled to the end of a transmission line. Maximum coupling occurs when the resonator is positioned centrally over the line end as in position 1. Position 2 indicates resonator positions for reduced coupling. Oltman has found empirically that for maximum power transfer from the line to the resonator and hence to radiation, for the resonator positioned centrally over the end, the ratio of the substrate thicknesses, T/t is about 2·7. Although not noted by Oltman, this ratio is expected to be a function of resonator and feed line width. Herein lies one of the particular advantages of the structures, that for a given element bandwidth determined by T and the element width w, the height of the feeding microstrip substrate is substantially less than for conventional coplanar feeding. This results in reduced radiation from the feed structure which was noted as a particular problem in certain antennas in Chapters 5 and 6. Alternately for a given feed substrate thickness, the radiating element height is increased resulting in an increased bandwidth. In addition to good matching capabilities for the centrally positioned resonator for the above thickness ratio, matching can be achieved for lower T/t ratios by moving the resonator parallel to the line. Oltman has suggested that an elliptical contour exists for the centre of the resonator for good match for $T/t < 2\cdot7$ as shown in Fig. 10.7(a). This then highlights a further advantage of this form, namely straightforward matching. Both electric and magnetic fields contribute in the arrangement of Fig. 10.7(*a*). However, when the resonator is orthogonal to the feed line as in Fig. 10.7(*b*) there will be little coupling between the magnetic fields and the resultant coupling will be capacitive. Here maximum coupling occurs when only the end of the resonator overlaps the line, as in 1; zero coupling occurs for central positioning, case 2.

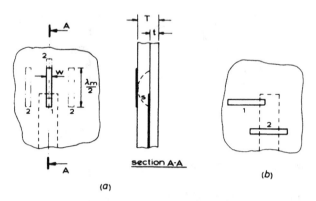

Fig. 10.7   *Overlaid electromagnetically coupled resonator radiator*
  1 — position for maximum coupling, 2 — position for reduced or zero coupling
  a Radiator coupled to microstrip feed line end
    S is elliptical contour of patch centre for good match
  b Radiator orthogonally coupled to microstrip feed line

Oltman has constructed an *X*-band array for radar tracking using this technique; a total of 128 elements are fed by corporate feeds in the four quadrants of a 280 mm diameter area and −13 dB sidelobes are achieved in both planes with losses, including the sum and difference circuitry, of 1·4 dB. A similar smaller array has also been described for the 14–16 GHz range with circularly polarised elements. Fig. 10.8 shows the radiating element details. A crossed resonator is used fed by two arms located below. A 90° phase delay is incorporated in one arm to generate the circular polarisation. The pill resistor on the splitter serves to isolate the two feed arms and improve the bandwidth of the circular polarisation as noted in Chapter 7; the ellipticity is less than 1·2 dB over the 14–16 GHz range.

Electromagnetic coupling has been used in a multiple layer microstrip antenna for multiple frequency operation (Sanford, 1978a). Fig. 10.9 shows a cut-away view of a three element example, where several resonant patches are mounted on top of each other. This is similar to the piggy-back concept except that only the top resonator is fed. At the frequency at which the top element is resonant, it will radiate with the next lower resonator as its ground plane. At the frequency of this lower resonator, the upper resonator forms an inductive coupling element that excites the lower element which then radiates. Many resonators can thus be placed on top of each other to form a multiple frequency antenna. The resonators can be either half-wavelength patches or quarter-wavelength shorted patches. It may be instructive to compare this antenna with the log periodic dipole array and the log periodic microstrip patch array, Fig. 10.11, described later. In these log periodic forms the higher frequency, non-resonant, elements form a loaded transmission line

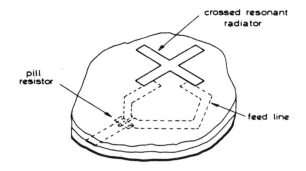

Fig. 10.8   *Circularly polarised overlaid resonator radiator*

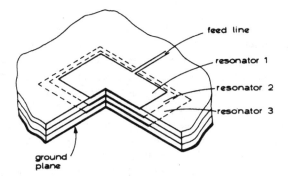

Fig. 10.9   *Multiple layer microstrip antenna*

which does not radiate; radiation is from a short active region. Similarly, non-resonant elements in the multiple layer antenna form a loaded transmission line to the active element or region. Radiation here is, as in the dipole array, towards the input end, whereas in the log periodic microstrip array radiation is normal to the direction of the transmission line. Although the multiple layer antenna is compact in required area, for wide bandwidth or many widely spaced frequencies of operation, the antenna height may be significant.

Another dual-frequency stacked resonator antenna has been described by Long and Walton (1979). The upper disc, shown in Fig. 10.10(a), is fed by a coaxial line, the inner of which passes through a clearance hole in the lower disc and is then connected to the upper disc at the appropriate point. Experiments were carried out in the 3 GHz region using two substrates 0·75 mm thick with $\epsilon_r = 2·47$. Separations of up to 10% between the two resonant frequencies were obtained for a diameter of 37·8 mm for the lower disc and between 37 and 38·5 mm for the upper disc. The radiation patterns are well behaved at both frequencies. It was noted by Long and Walton that dual frequency behaviour was not obtained when the lower disc was driven with the upper one acting as a parasitic.

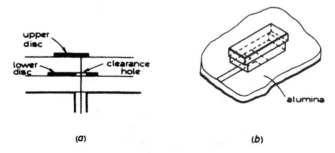

(a)   (b)

Fig. 10.10   *Stacked microstrip antennas*
a Dual-frequency stacked antenna
b Wideband stacked antenna for circuit integration

Hall *et al.* (1979) have shown that wide bandwidth operation can be obtained when the lower element is fed using the structure shown in Fig. 10.10(*b*). Here rectangular half-wavelength resonators are used and the lower substrate is alumina ($\epsilon_r \approx 9.8$), 0·625 mm thick. The resonator dimensions were determined empirically to give closely spaced resonances such that continuous use over a broad bandwidth is possible. For the 10 GHz three layer example shown, the lower resonator was 4 × 4·6 mm and the middle and upper resonators etched on 1·59 mm thick substrate were 4·5 × 7·8 mm and 4·5 × 8·5 mm, respectively. A bandwidth of 18% was achieved for a 2:1 input VSWR and the antenna had a measured gain of greater than 5·3 dB. In addition to broad bandwidth, the use of alumina substrates also allows direct integration with other radio frequency microstrip circuitry. Although the dielectric was cut off flush with the upper resonators for convenience in design, antennas with similar characteristics have also been produced using continuous substrates to produce a rugged structure.

The overlaid resonator technique has been used to form a log periodic microstrip patch array antenna that has wide bandwidth capabilities (Hall, 1980). The structure is shown in Fig. 10.11(*a*). The patch resonators lie transverse to the line and are displaced on alternate sides of the line to produce a broadside beam with $\lambda_m/2$ spacing. The dimensions and spacing of the resonators are increased in a log periodic manner along the array analogous to the log periodic free-space dipole array (Isbell, 1960), Fig. 10.11(*b*). In both arrays, nonresonant high-frequency elements capacitively load the transmission line, but do not radiate. Elements near resonance radiate strongly and form an 'active' region. Those beyond this region, that are resonant at lower frequencies, do not see significant power and do not contribute to radiation. A 9-element example of the microstrip patch array using upper and lower substrates 0·793 mm thick, with $\epsilon_r = 2\cdot32$, had a bandwidth of 8–10·75 GHz (30%) for 2·2:1 VSWR with gain greater than 6·5 dB and better than 70% efficiency. It should be noted that although this form has considerably wider bandwidth than the stacked resonator form it is about $3\lambda_0$ long at midfrequency. However the various options may allow trade-offs between array length and height to be made to suit the particular application.

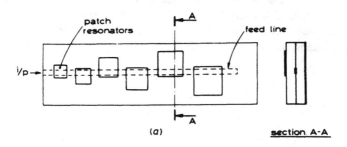

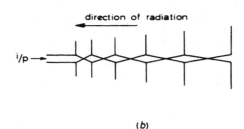

Fig. 10.11   *Log-periodic antennas*
a Log-periodic microstrip patch array
b Log-periodic free-space dipole array

10.2.3 Low-profile dipole antennas

For conventional operation, dipoles are located either in free space or spaced $\lambda/4$ above a ground plane. If the ground plane to dipole spacing is reduced, then the input resistance of the element falls significantly, reducing the efficiency of radiation. However, there are techniques available for improving this, such as the use of broadband matching networks or by folding the dipole, that can be used and such techniques allow an element to be developed that is similar to a microstrip radiator in that its profile is small and it is amenable to printed-circuit type production. In addition, microstrip feed networks can be used that allow straight-forward construction of the matching network. In this Section then, the low-profile dipole element is examined and compared to microstrip equivalents.

The dipole is shown in Fig. 10.12. Developed initially by Sidford (1973) and incorporated into a dipole-slot configuration for circular polarisation, much work on the element has since been done by Dubost, (Dubost and Havot, 1975; Dubost *et al.*, 1976; and Dubost, 1978). It can be seen that the folded dipole is printed on the underside of a thin sheet of dielectric. The microstrip feed network is printed on the upper side and connected to one side of the gap in the dipole by a shorting pin (Sidford, 1973) or by capacitive coupling (Dubost *et al.*, 1976). The coaxial feed line outer is connected to the folded arm of the dipole at the centre and the inner is connected to the microstrip line. Thus the dipole element forms the microstrip ground plane. The whole is then suspended over the ground plane with typically $0.01 < h/\lambda_0 < 0.1$. By the use of a matching stub, Sidford was able to match this element over a bandwidth of about $\pm 5\%$ for $2:1$ VSWR at 1.55 GHz for $h/\lambda_0 \approx 0.08$. Dubost has obtained similar results.

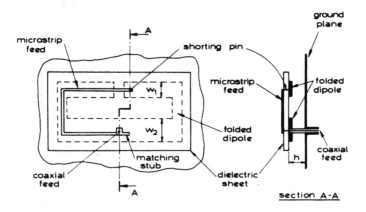

Fig. 10.12   *Folded dipole element close to ground plane*

Some understanding of the performance of such elements can be obtained by examining the methods of dipole matching mentioned above. The radiation resistance of a dipole located close to a ground plane has been calculated by Dubost *et al.* (1976) and some results are shown in Table 10.1. This low input resistance on resonance can be matched in two ways. First, if dipole folding is used, then the expression for the impedance transformation ratio, T_z, for a dipole in free space, can be used to give some indication of the improvement obtained for dipoles close to the ground plane. T_z is given by Jasik (1961a) as:

$$T_z = \frac{Z_f}{Z_r} = \left[1 + \frac{\cosh^{-1}\left(\dfrac{a^2 - b^2 + 1}{2a}\right)}{\cosh^{-1}\left(\dfrac{a^2 + b^2 - 1}{2ab}\right)} \right]^2 \tag{10.1}$$

where Z_f and Z_r are the input impedances of the folded and isolated dipoles, respectively, $a = d/r_1$ and $b = r_2/r_1$ where d, r_1 and r_2 are the spacing, and radii

Table 10.1 *Radiation resistance of dipole spaced h/λ_0 from conducting ground-plane*

h/λ_0	0·1	0·05	0·02	0·005
$R(\Omega)$	23·4	5·84	0·94	0·06

Table 10.2 *Bandwidths of microstrip patches and folded dipoles close to ground-plane*

h/λ_0	B, bandwidth (%) for 2:1 VSWR		Dielectric ϵ_r	Dipole matching network
	folded dipole	microstrip patch		
0·0175	5·0	3·3	1 (air)	Series and parallel capacitances
0·019	5·9	1·6	2·3	Series and parallel capacitance and high Z line
0·084	17·5	6·8	2·3	Series and parallel capacitances

of the fed and folded dipole arms, respectively. If an equivalent radius, r_e, is used in place of the strip widths w_1 and w_2, given by (Jasik, 1961a) as

$$r_e = 0.25 w_{1,2} \tag{10.2}$$

then for Sidford's dipole, $a \approx 3$ and $b \approx 2$ giving $Z_f/Z_r \approx 8$. This will then produce an input resistance greater than 50Ω; hence a small final matching element only will be necessary. In Dubost's dipole, generally $w_1 > w_2$ giving impedance transformation ratios much less than 4. In this case matching is obtained using series and parallel tuned circuits as shown in Fig. 10.13 that also provide some broadbanding of the input match (Jasik, 1961b). Using such matching networks Dubost has achieved bandwidths at 2·2 GHz, as shown in Table 10.2. Also shown are bandwidths, B, for square microstrip patches calculated from eqn. 4.48 of Chapter 4, which gives for a 2:1 VSWR,

$$B = 188 \cdot \frac{h}{\epsilon_r \lambda_0} \% \tag{10.3}$$

It is clearly seen that for an air dielectric, the folded dipole is about 50% broader in bandwidth than a square microstrip patch. Dubost (1979) has shown for the short-circuited half dipole that the broadbanding network increases the bandwidth by a factor of about 40%, suggesting that the uncompensated full dipole element and microstrip patch, both with air spacing, have equivalent bandwidths. However, when dielectric is introduced, the heavy-wave trapping action of the microstrip patch gives significantly decreased bandwidth whereas the dipole element is substantially unchanged. It is believed that this improved bandwidth capability arises from the fact that the dipole is fed against itself with the ground plane free to find its own potential whereas in the microstrip case the patch is fed against the ground

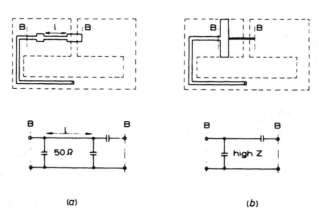

(a)   (b)

Fig. 10.13   *Matching networks for folded dipoles*
   a Series and parallel capacitances
   b Series and parallel capacity and high-impedance line

plane. This difference in feeding also highlights one of the main differences between the dipole element and the patch and that is complexity of construction. The microstrip patch is made using one etching operation whereas the dipole element needs two etching operations with good alignment of the patterns on each side of the board, which is then mounted above the ground plane. In other areas of electrical performance the two forms are similar with values quoted by Dubost *et al.* (1976) of gain about 7 dB, beamwidths about 60 x 75° and cross-polarisation less than − 20 dB.

Dubost (1979) has also described a short circuited microstrip dipole. Shown in Fig. 10.14, the length of the element l is $0.2\lambda_0$, width w is $0.23\lambda_0$ and height h is $0.05\lambda_0$. The dielectric sheet thickness is $0.0002\lambda_0$. A metal plate forms the short-circuiting plane along one edge of the element. The microstrip feed circuit consists of two short-circuited compensating lines, C, and a high-impedance section, Z. The high-impedance line is gap coupled to the radiating element and at its end is short-circuited to the ground plane. This short circuit means that the element is driven against the ground plane in contrast to the folded dipole elements described earlier. Thus the element is similar to a microstrip short-circuited quarter-wavelength patch as described in Chapter 4, with the addition of an overlaid feed circuit. This similarity is endorsed when the bandwidth of the element without matching elements, designed for operation at 0.73 GHz, is compared to a microstrip short-circuited quarter-wavelength patch in Table 10.3, where comparable bandwidths are evident. Also shown is the bandwidth of the dipole element with the compensation

Table 10.3   *Bandwidths of short circuited air spaced dipole and microstrip elements*

	Bandwidth % for 2:1 VSWR	
	Short-circuited half dipole	Short-circuited quarter-wavelength microstrip patch
Without reactive compensation (theoretical)	8	9
With compensation (measured)	12	–

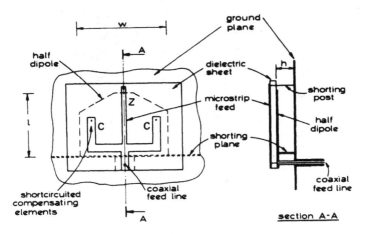

Fig. 10.14   *Short-circuited half-dipole element*

added and a significant increase is noted. A linear array of 32 such elements (Dubost and Vinatier, 1980), fed by a microstrip corporate feed was found to have sidelobes of about −15 dB for a uniform distribution. However, radiation from the feed network was found to be excessive and gave rise to high far out co- and cross-polarised sidelobes.

10.3 Antenna feed structures in microstrip

10.3.1 Feeds for surface-wave structures

A microstrip disc antenna has been used to feed a circular corrugated surface (Bailey, 1977) as shown in Fig. 10.15. Designed for 5·4 GHz, the disc is fed at two points with phase quadrature to achieve circular polarisation and is located flush with the corrugated surface which has a corrugation depth and width of $0·343\lambda_0$ and $0·114\lambda_0$, respectively. The effect of the corrugations is to equalise the pattern beamwidths and improve the ellipticity at low elevation angles. The antenna was intended for a microwave landing aid for helicopters which approach at low angles and land on top of the antenna. Good ellipticity at low angles helps to minimise the specular multipatch effects in the vicinity of buildings. The measured patterns showed an ellipticity of less than 5 dB down to a 10° elevation angle with a gain of 7 dB in the vertical direction.

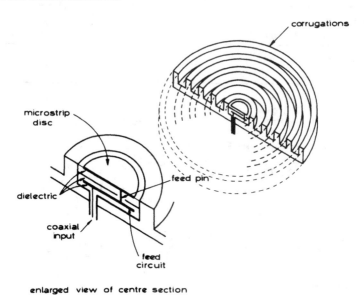

enlarged view of centre section

Fig. 10.15   *Sectional view of multistrip disc-fed corrugated antenna*

Kerr (1979) has described a Yagi array using a microstrip patch as the feed element. Some improvements in the radiation pattern are noted particularly in the backward direction.

10.3.2 Feeds for reflector configurations

An integrated feed array and comparator for a Cassegrain type reflector for a 94 GHz monopulse guidance radar is described by Oltman *et al.* (1978). The antenna configuration is shown in Fig. 10.16(*a*). The microstrip array is located at the prime focus of the reflector system the whole of which is mounted on gimbals to provide a scanning capability. The choice of the 94 GHz band for this application is based on improved directivity and adverse weather performance obtainable in comparison to lower frequency systems. However, construction of microstrip circuits at this frequency involves difficulties, in particular with connectors and the practicability of empirical development. The latter problem was solved by developing the circuit using a 28·4:1 scale model. For a fused silica ($\epsilon_r = 3 \cdot 83$) substrate 0·11 mm thick at 94 GHz, the scale model used a 3·1 mm thick substrate at 3·31 GHz. The circuit was optimised using copper tape conductors. The silhouette was then photographically reduced and used for the 94 GHz production. The array and comparator layout is shown in Fig. 10.16(*b*). Four half-wavelength patches are fed at 50Ω points and connected to the four branch line couplers, the outputs 1–4 of which provide the sum and difference patterns in each plane. The curved feed lines are used to reduce unwanted radiation. The scale model had well behaved radiation patterns with a

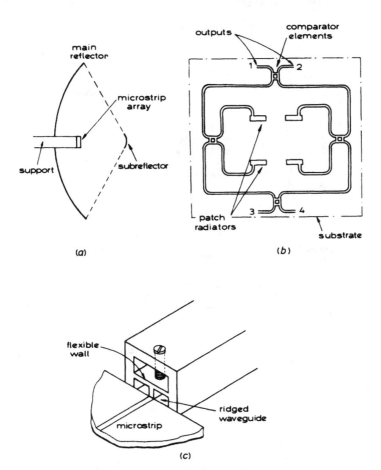

Fig. 10.16   *Cassegrain reflector antenna with microstrip feed*
    a Schematic diagram of antenna system
    b Microstrip array and comparator layout
    c 94 GHz waveguide to microstrip transition used in antenna development

30 dB null in the difference pattern. Measurements on the 94 GHz circuit were made with microstrip to waveguide transitions as shown in Fig. 10.16(*c*). Here a ridge waveguide is matched to a plane waveguide by a 5-section Chebyshev taper. The flexible upper wall allows the ridge to be pressed down onto the microstrip conductor. This transition had a VSWR of 1·23:1 over the 90 to 100 GHz range. Measurements at 94 GHz indicate that the losses in the array and comparator circuit

will be of the order of 0·8 dB and developments will involve the integration of mixers on to this circuit board to allow IF connections only to be made to the assembly.

Kerr (1979) has used a single microstrip disc to feed a reflector antenna operating at 1·25 GHz. The single coaxial feed was found to give significantly less

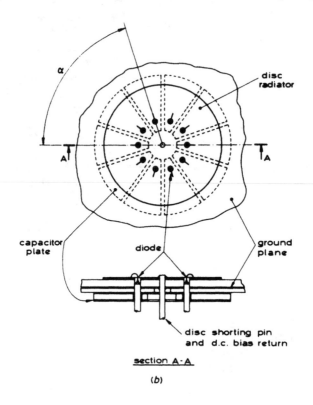

section A·A

(*b*)

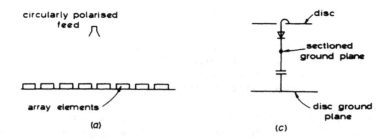

Fig. 10.17   *Reflectarray using microstrip discs*
    a Reflectarray configuration
    b Switch microstrip disc element
    c Diode RF circuit

radiation pattern degradation than a tripod support or a waveguide shepherd's crook feed.

A disc element has been designed for use in a reflect-array configuration (Malagisi, 1977; Montgomery, 1978). The configuration is shown in Fig. 10.17(*a*). The reflector surface is formed from a planar array of disc elements. The array is excited by a circularly polarised feed located at a distance sufficient to adequately illuminate the array. The resultant beam is formed and scanned by altering the phase of the wave reflected from each element. The performance of the complete system is not given but a similar reflect array, using cavity backed spiral antennas as the array elements, is described by Phelan (1976, 1977) and scanning out to ± 60° is readily achieved. In this case, four phase states are achieved using a four-arm spiral element backed by the diode switches and drive circuits. In the microstrip disc case a variable phase reflection coefficient is obtained by short-circuiting it at various angular positions by means of diodes mounted in the disc, Fig. 10.17(*b*). The disc RF circuit is shown in Fig. 10.17(*c*). The disc centre is shorted with a post that forms the d.c. return path. The diodes are mounted on posts that penetrate but do not contact the ground plane. The RF circuit is completed by a sectored plane mounted close to the ground plane that produces capacitive coupling. By shorting out pairs of diodes diagonally opposite, a phase shift of twice the angle of displacement of the diodes from a reference plane, α in Fig. 10.17(*b*), in the reflected wave is achieved. The primary reflection loss is caused by losses in the dielectric which for the 3 GHz disc measured was 0·2 dB and diode losses which are less than 0·4 dB. Usable bandwidths are determined by the reflected wave depolarisation and are typically 1%.

10.3.3 Slot antenna feeds
The microstrip line can be used as a convenient feed structure for antennas formed from slots cut in the ground plane. The resultant antenna is then either used in an omnidirectional mode, radiating on both sides of the ground plane, or backed by a reflecting sheet $\lambda_0/4$ away to produce a single main beam. The single narrow ground plane slot fed by microstrip is described by Nakaoka *et al.* (1977) and is shown in Fig. 10.18. Curves of radiation resistance and reactance calculated using a variational method are compared to measurements. A square array of 16 such elements (Yoshimura, 1972) having a Dolph-Chebyshev aperture distribution had -24 and -22 dB sidelobes in the *H*- and *E*-planes, respectively, at *X*-band.

A 512-element array has been produced (Collier, 1977) for a 12 GHz TV reception application. The slots were made on a 0·79 mm thick substrate and mounted $\lambda_0/4$ in front of a large ground plane. Sidelobes higher than -6 dB measured in the *E*-plane were thought to be due to phase errors in the feed structure; conductor losses in the feed structure shown in Fig. 6.21 are about 5 dB with an additional 2 dB due to other feed losses at the bends and splitters. The resultant array gain is > 23 dB, over a 14% bandwidth.

Other slot antennas, such as the spiral (Van de Capelle *et al.*, 1979), have been fed by microstrip. In this case, performance is similar to that obtained using other feed methods.

10.4 Conformal microstrip antennas

The basic microstrip patch radiator can be used in various applications where a flat radiator is inappropriate and the shape of the mounting surface dictates the radiating aperture. The most commonly encountered surface is the cylinder and several forms have been developed to produce broad beam radiation patterns when mounted on such a surface.

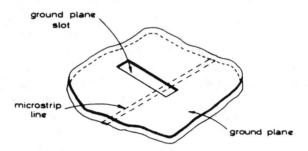

Fig. 10.18   *Ground plane slot fed by microstrip line*

The first form is the spiral-slot (Sindoris *et al.*, 1978). A short-circuited quarter-wavelength microstrip patch is wrapped round the cylindrical surface to form a spiral. The development of the idea is illustrated in Fig. 10.19 where Fig. 10.19(*a*) shows the quarter-wavelength shorted patch. In Fig. 10.19(*b*), the patch is titled by α to allow it to be wrapped around the cylinder as in Fig. 10.19(*c*). The cylinder is an epoxy fibreglass dielectric and the copper conductors are added using an electroless plating, masking and electroplating technique (Jones, 1974). The lower end and the inside of the patch are similarly plated to form the short circuit and ground plane. The spiral slot has a height and diameter of $0.06\lambda_0$ but unlike conventional small antennas has a well matched input impedance of less than $2:1$ VSWR over a 2% bandwidth at 238 MHz. The radiation patterns are similar to those of a dipole oriented parallel to the cylinder axis and the $+1$ dB gain indicates an efficiency of better than 50%. The antenna was developed for mounting in a small diameter missile; when mounted in the missile some lobing in the radiation pattern is noted due to excitation of the missile body. The spiral slot has also been developed for 42 MHz application (Krall *et al.*, 1979) in which the antenna had to be contained in a $0.04\lambda_0 \times 0.15\lambda_0$ cylindrical volume. The tube and hence the microstrip resonator thickness was $0.001\lambda_0$ which, together with a dielectric loss tangent of 0.01, resulted in an antenna with a 1% efficiency and bandwidth. It should be noted that the tolerance control of the resonator dimensions and dielectric constant is not sufficient to allow repeatable resonant frequencies to be achieved; trimming back the resonator at points T_1 and T_2, increase and decrease, respectively, the resonant frequency and this is used for hand tuning of each manufactured antenna.

An element for a similar application is the microstrip wraparound antenna (Munson, 1974) and is shown in Fig. 10.20. The radiator is a very wide patch fed by a corporate network and the electrical design is described in Section 5.2.1 of Chapter 5.

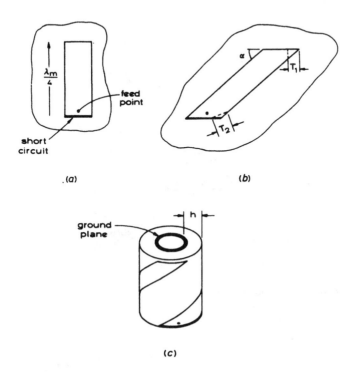

(a)　　　　　　　　　　　　　　　　(b)

(c)

Fig. 10.19　*Microstrip spiral slot radiator*
　　a Short-circuited quarter-wavelength patch
　　b Patch tilted by angle α
　　c Tilted patch wound round cylinder to form spiral slot

Another microstrip antenna with a cylindrical aperture is the edge-slot radiator (Jones, 1978) which is shown in Fig. 10.21(*a*). Here a cylindrical cavity is formed by plating the top and bottom of a dielectric slab whilst the edge is left open. The radiator is fed at the centre by a coaxial line; the outer of the line is connected to one plate and the inner is connected to the other. The dominant mode is a radial one (Sengupta *et al.,* 1979) and radiation is circularly symmetric in the plane of the disc. By suitable positioning of shorting posts the radiator can be tuned within a six to one frequency range. Three of these have been coupled in series to form a multiple frequency antenna (Schaubert and Jones, 1979) as shown in Fig. 10.21(*b*), operating at 625, 790 and 875 MHz. The antenna was mounted in a long cylinder similar to the spiral slot application and whilst good azimuthal symmetry was maintained, lobing of the elevation pattern occurred due to excitation of the cylinder.

The parallel plate edge slot configuration has also been used to form an antenna for a hemispherical body (Jones, 1978) as shown in Fig. 10.22. Here two orthogonally positioned edge slot radiators are fed independently in any phase relationship to vary the direction and polarisation of the radiated field. The example quoted was for use at 560 MHz.

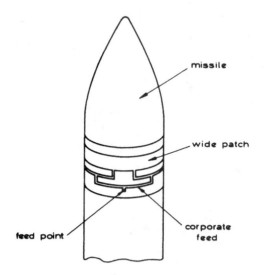

Fig. 10.20   *Wraparound microstrip antenna*

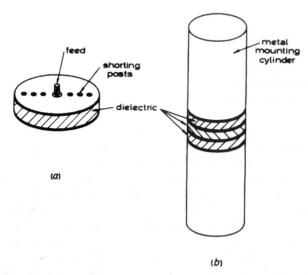

Fig. 10.21   *Edge slot radiator*
   a Single element
   b Travelling-wave array

A further example of a conformal patch radiator is the dual frequency antenna (Sanford and Munson, 1975) noted in Section 10.1.1 and shown in Fig. 10.4. The patch was mounted on the outer surface of a spacecraft and was curved to conform to the surface of the vehicle. No details are given on the radius of curvature involved but it is noted that the radiation patterns obtained are similar to those obtained from a flat patch.

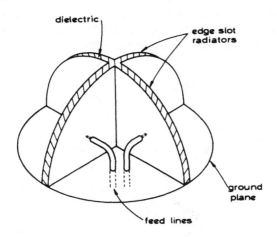

Fig. 10.22   *Hemispherical edge slot radiator*

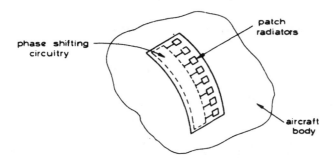

Fig. 10.23   *Conformal array for aircraft application*

A patch array designed for an aircraft to satellite communication link is described by Sanford (1978b) and shown in Fig. 10.23. Eight patches are mounted on the same substrate together with phase shifting and feeding circuitry to scan the beam in the elevation direction. Designed for operation at 1·5 GHz the array including radome is 3·6 mm thick. Element-phasing was optimised for maximum multipath rejection at low scan angles and to account for the curvature of the mounting surface. An improved microstrip radiator for a similar application is described by Hall *et al.* (1981). The element, shown in Fig. 1.1, has in addition to improved bandwidth enhanced gain to circular polarisation at low elevation angles when compared to a conventional rectangular patch.

A microstrip array mounted on a cylinder and designed to produce a conical beam along the cylinder axis is described by Hall *et al.* (1981) and is shown in Fig. 10.24. The array consisted of three rampart line arrays equispaced round the cylinder and parallel to its axis. Less than 3 dB ripple was obtained in the ϕ plane in the main beam whilst sidelobes in a conical volume along the cylinder axis but opposite the main beam were around -12 dB. The high sidelobe level is attributed to enhanced coupling to the surface substrate mode for end fire microstrip arrays and radiation from the microstrip to coaxial transitions.

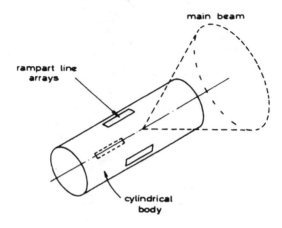

Fig. 10.24   *Conical beam microstrip array for cylindrical body*

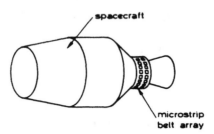

Fig. 10.25   *Satellite microstrip belt array*

A belt array of microstrip patches has been used for a telemetry and command link for a satellite (Ostwald and Garvin, 1975), shown diagrammatically in Fig. 10.25. Two rows of 64 circularly polarised elements designed for 2 GHz operation are used. All elements around the belt are fed in phase with equal amplitude to provide an omnidirectional pattern in this plane; a ripple of less than 5 dB was measured. The environment encountered by such satellite antennas is particularly severe and Ostwald quotes limits of $+97°C$ to $-144°C$ for this application and notes that control of these temperatures is obtained by the use of radiative finishes applied to the antenna surfaces. The effect of temperature on patch antenna performance is discussed further in Chapter 8.

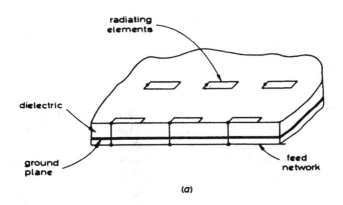

(a)

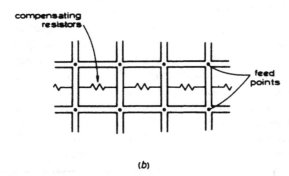

(b)

Fig. 10.26 *Spherical microstrip array*
a Array configuration
b Feed network

An antenna for mounting on a 2 m radius sphere has been constructed by Dubost *et al.* (1972). A section through the array is shown in Fig. 10.26(*a*). The radiating elements are located on the upper substrate which is 3 mm thick and is a mixture of small glass spheres in an epoxy resin. The feed network, shown in Fig. 10.26(*b*), is etched on a 0·25 mm thick substrate and consists of a lattice network with feed points at each intersection. Compensating resistors are located as shown and serve to give $-13\,dB$ sidelobes between 8·8 and 9·8 GHz with beamwidths between 5 and 8 degrees; the resistors, however, reduce the overall efficiency to less than 4%.

10.5 The future

The microstrip antenna was first suggested in 1955 but for reasons discussed in Chapter 1 did not receive wide attention until about 1970. Since then there has been a large increase in the literature on microstrip antennas both in terms of theoretical approaches and experimental designs. The copious literature now available indicates that microstrip antennas may soon be widely applied in practical systems; however, it appears that few microstrip antennas have so far emerged in manufactured equipment and their future application is still a somewhat open question. A few comments about the future are therefore appropriate in this closing Chapter.

This book has established that the advantages of low cost, flat profile and integration potential of these antennas is impaired, at present, by low efficiency or narrow bandwidth, combined with significant tolerance problems. In some applications the performance so far obtained is adequate, but in others the poorer electrical performance has probably deterred designers from taking immediate advantage of the considerable mechanical and constructional benefits. One is then led to ask whether this trend will continue into the future whereby microstrip antennas remain a direct alternative to other forms, so that the system designer will be faced with a straight choice between a cheaper, lower performance microstrip antenna or a more expensive, higher performance, conventional type. Or alternatively, will the trend be towards combining the advantages of microstrip and conventional forms to produce hybrid arrangements, some of which have been discussed in this chapter. For example the multilayer configuration combines the advantages of microstrip feed lines with bandwidth improvements offered by thick substrates, while the reflector/microstrip feed configuration combines the integration possibilities of microstrip antennas and circuits with the efficiency and control of a reflector antenna design. This trend appears to be mirrored in everyday commodities where cheaper materials are mixed with more expensive ones in an optimal hybrid manner; for example, the throw-away ball-point pen combines the precision of construction in metal at the writing tip with a cheap plastic body.

Readers may well be aware of other noticeable trends with microstrip antennas and without doubt their future will be greatly influenced by future system requirements, both military and commercial. One aspect is however very clear: the flat thin profile of microstrip antennas is an advantage of immense importance and it will continue to greatly influence the decisions made by system designers for many years to come and inspire continued innovation in the antenna field.

10.6 References

BAILEY, M. C. (1977): 'A broad beam circularly-polarised antenna'. IEEE Ap-S Symposium, Stanford, USA

COLLIER, M. (1977): 'Microstrip antenna array for 12 GHz TV', *Microwave J.*, Sept, pp. 67–71

DUBOST, G., BIZOUARD, A., BEAULIEU, L., and RANNOU, J. (1972): 'Antenne de reseau de surface a large bande de frequence', *Revue Technique Thomson-CSF*, 4, p. 333

DUBOST, G., and HAVOT, H. (1975): 'Doublet replie en plaques'. Brevet d'invention PV No 75.151.14 du 15.05.1975

DUBOST, G., NICOLAS, M., and HAVOT, H. (1976): 'Theory and applications of broadband microstrip antennas'. Proc. 6th European Microwave Conference, Rome, pp. 275–279

DUBOST, G. (1978): 'Bandwidth of a dipole near and parallel to a conducting plane', *Electron. Lett.*, 14, pp. 734–736

DUBOST, G. (1979): 'Theory and experiments of a broadband short circuited microstrip dipole at resonance'. Proc. workshop on printed antenna technology, New Mexico, USA, pp. 32.1–32.13

DUBOST, G., and VINATIER, C. (1980): 'High gain array at 12 GHz for telecommunications'. Int. URSI Symp. on Electromagnetic Waves, Munich, pp. 215 A/1 to A/4

GREISER, J. W. (1976): 'Coplanar stripline antenna', *Microwave J.*, Oct., pp. 47–49

HALL, P. S., WOOD, C., and GARRETT, C. (1979): 'Wide bandwidth microstrip antennas for circuit integration', *Electron. Lett.*, 15, pp. 458–460

HALL, P. S. (1980): New wideband microstrip antenna using log-periodic technique', *Electron. Lett.*, 16, pp. 127–128

HALL, P. S., WOOD, C., and JAMES, J. R. (1981): 'Recent examples of conformal microstrip antenna arrays for aerospace applications'. IEE Int. Conf. on Antennas and Propagation, York, England, pp. 397–401

ISBELL, D. E. (1960): 'Log periodic dipole arrays', *IRE Trans.*, AP-8, pp. 260–267

JASIK, H. (1961): 'Antenna engineering handbook' (McGraw-Hill, New York) (*a*) sect. 3, (*b*) p. 31-27

JONES, H. S. (1974): 'Design of dielectric-loaded circumferential slot antennas of arbitrary size for conical and cylindrical bodies', Report HDL-TR-1684, Harry Diamond Laboratories, Adelphi, Maryland, USA

JONES, H. S. (1978): 'Some novel design techniques for conformal antennas'. Proc. IEE Int. Conf. on Ant. and Prop., London, pp. 448–452

KALOI, C. (1976): 'Notch fed electric microstrip dipole antenna'. US Patent 3,947,850

KERR, J. (1977): 'Other microstrip antenna applications'. Proc. 1977 Antenna Applications Symposium, Illinois, USA

KERR, J. (1979): 'Microstrip antenna developments'. Proc. workshop on printed circuit antenna technology, New Mexico, USA, pp. 3-1 to 3-20

KRALL, A. D., McCORKLE, J. M., SCARZELLO, J. F., and SYELES, A. M. (1979): 'The omni microstrip antenna: a new small antenna', *IEEE Trans.*, AP-27, pp. 850–853

LONG, S. A., and WALTON, M. D. (1979): 'A dual frequency stacked circular disc antenna', *IEEE Trans.*, AP-27, pp. 270–273

MALAGISI, C. (1977): 'Electrically scanned microstrip antenna array'. US Patent 4,053,895

McILVENNA, J., and KERNWEIS, N. (1979): 'Modified circular microstrip antenna elements', *Electron. Lett.*, 15, pp. 207–208

MONTGOMERY, J. P. (1978): 'A microstrip reflect array antenna element'. Proc. 1978 Antenna Applications Symposium, Illinois, USA

MUNSON, R. E. (1974): 'Conformal microstrip antennas and microstrip phased arrays', *IEEE Trans.*, AP-22, pp. 74–78

NAKAOKA, K., ITOH, K., and MATSUMOTO, T. (1977): 'Input characteristics of slot antenna for printed array antenna', *Trans. IECE Japan*, E60, pp. 256–257

OLTMAN, H. G. (1977): 'Microstrip dipole antenna elements and arrays thereof'. US Patent 4,054,874

OLTMAN, H. G. (1978): 'Electromagnetically coupled microstrip dipole antenna elements'. Proc. 8th European Microwave Conference, Paris, pp. 281–285

OLTMAN, H. G., WEEMS, D. M., LINDGREN, G. M., and WALTON, F. D. (1978): 'Microstrip components for low cost millimeter wave missile seekers', *in* 'Millimeter and submillimeter wave propagation and circuits'. AGARD Conf. Proc. 245, Munich, Germany, pp. 27-1 to 27-9

OSTWALD, L. T., and GARVIN, C. W. (1975): 'Microstrip command and telemetry antennas for communications technology satellite' *in* 'Antennas for aircraft and spacecraft'. IEE Conf. Publ. 128, pp. 217–222

PHELAN, H. R. (1976): 'Spiraphase – a new, low cost, lightweight phased array', *Microwave J.*, 19, pp. 41–44

PHELAN, H. R. (1977): 'Dual polarised spiraphase', *Microwave J.*, 20, pp. 37–40

SANFORD, G. G., and MUNSON, R. E. (1975): 'Conformal VHF antenna for the Apollo-Soyuz test project' *in* 'Antennas for aircraft and spacecraft'. IEE Conf. Publ. 128, pp. 130–135

SANFORD, G. G. (1978a): 'Multiple resonance radio frequency microstrip antenna structure'. US Patent 4,070,676

SANFORD, G. G. (1978b): 'Conformal microstrip phased array for aircraft tests with ATS-6', *IEEE Trans.*, AP-26, pp. 642–646

SENGUPTA, D. L., MARTINS-CAMELO, L. F., JONES, H. S., and SCHAUBERT, D. H. (1979): 'Theory of the input behaviour of a dielectric-filled edge-slot antenna'. IEEE AP-S, Int. Symp. Dig., Washington, USA, pp. 138–141

SCHAUBERT, D. H., and JONES, H. S. (1979): 'Series fed, dielectric filled, edge slot antennas'. IEEE AP-S, Int. Symp. Dig., Washington, USA, pp. 142–145

SIDFORD, M. J. (1973): 'A radiating element giving circularly polarised radiation over a large solid angle'. IEE Conf. Publ. 95, pp. 18–25

SINDORIS, A. R., SCHAUBERT, D. H., and FARRAR, F. G. (1978): 'The spiral slot – a unique microstrip antenna'. Proc. IEE Int. Conf. on Ant. and Prop., London, pp. 150–154

VAN DE CAPELLE, A., DE BRUYNE, J., VERSTRAETE, M., PUES, H., and VANDENSANDE, J. (1979): 'Microstrip spiral antennas'. Proc. IEEE Int. Symp. on Ant. and Prop., Seattle, USA, pp. 383–386

WEN, C. P. (1969): 'Coplanar waveguide: a surface strip transmission line suitable for non-reciprocal gyromagnetic device applications', *IEEE Trans.*, MTT-17, pp. 1087–1090

YOSHIMURA, Y. (1972): 'A microstripline slot antenna', *IEEE Trans.*, MTT-20, pp. 760–762

Appendixes

Appendix A

Values of microstrip impedance and effective relative permittivity computed by programme IPED which is based on eqns. 2.4, 2.5 and 2.7

ϵ_r	h mm	w mm	Z_m Ω	ϵ_e frequency = 0 Hz	ϵ_e frequency = 10 GHz
		0·5	115·8	1·742	1·767
2·3	0·79	1·0	85·1	1·843	1·874
		2·0	55·5	1·918	1·963
		0·25	101·0	2·725	2·754
4	0·5	0·5	75·5	2·896	2·933
		1·0	52·2	3·055	3·108
		0·25	66·6	6·263	6·411
10	0·5	0·5	49·5	6·690	6·890
		1·0	33·7	7·165	7·471

Appendix B[†]

Closed-form empirical formulas for microstrip discontinuities
Garg *et al.* (1978) modify an expression by Hammerstad and Bekkadal (1975) for the end effect Δl of an open-circuit microstrip line, Fig. 2.9(*a*), giving

$$\frac{\Delta l}{h} = 0.412 \left(\frac{\epsilon_e + 0.3}{\epsilon_e - 0.258} \right) \left[\frac{w/h + 0.264}{w/h + 0.8} \right] \tag{A.1}$$

for $w/h \geqslant 0.2$ and $2 \leqslant \epsilon_r \leqslant 50$.

The microstrip gap, Fig. 2.9(*b*), is characterised by capacitances C_p and C_g where

$$C_{\text{even}} = 2C_p$$
$$C_{\text{odd}} = 2C_g + C_p \tag{B.2}$$

[†] References refer to those of Chapter 2.

Garg *et al.* (1978) give the expressions

$$C_{odd}/w\,(pF/m) = \left(\frac{s}{w}\right)^{m_{od}} \exp(k_{od})$$

$$C_{even}/w\,(pF/m) = \left(\frac{s}{w}\right)^{m_{ev}} \exp(k_{ev})$$

(B.3)

where

$$m_{od} = \frac{w}{h}(0.619 \ln(w/h) - 0.3853)$$

$$k_{od} = 4.26 - 1.453 \ln(w/h) \qquad\qquad 0.1 \leqslant s/w \leqslant 1.0$$

$$m_{ev} = 0.8675, \quad k_{ev} = 2.043\left(\frac{w}{h}\right)^{0.12}; \qquad 0.1 \leqslant s/w \leqslant 0.3$$

$$m_{ev} = \frac{1.565}{(w/h)^{0.16}} - 1, \quad k_{ev} = 1.97 - \frac{0.03}{w/h}; \qquad 0.3 \leqslant s/w \leqslant 1.0$$

(B.4)

eqns. B.3 and B.4 apply to the strip width range $0.5 \leqslant w/h \leqslant 2$ and $\epsilon_r = 9.6$, but the capacitances for other ϵ_r values in the range $2.5 \leqslant \epsilon_r \leqslant 15$ are obtained by scaling the results, thus

$$C_{odd}(\epsilon_r) = C_{odd}(9.6)(\epsilon_r/9.6)^{0.8}$$

$$C_{even}(\epsilon_r) = C_{even}(9.6)(\epsilon_r/9.6)^{0.9}$$

(B.5)

No information is given about ϵ_r values in the range $2 < \epsilon_r < 2.5$.

For the step discontinuity, Fig. 2.9(c), Garg *et al.* (1978) and Gupta *et al.* (1979) give the following equivalent circuit parameters:

$$L_1 = \frac{L_{w1}}{L_{w1} + L_{w2}} \cdot L_s; \qquad L_2 = \frac{L_{w2}}{L_{w1} + L_{w2}} \cdot L_s$$

$$L_{w\frac{1}{2}} = Z_{m\frac{1}{2}}(\epsilon_{e\frac{1}{2}})^{1/2}/c\,(H/m)$$

(B.6)

where $c = 3 \times 10^8$ m/s and $Z_{m\frac{1}{2}}$ and $\epsilon_{e\frac{1}{2}}$ are values of Z_m and ϵ_e appropriate to the widths w_1 and w_2, respectively, and

$$\frac{L_s}{h}\,(nH/m) = 40.5\left(\frac{w_1}{w_2} - 1.0\right) - 75\ln\left(\frac{w_1}{w_2}\right) + 0.2\left(\frac{w_1}{w_2} - 1.0\right)^2 \quad (B.7)$$

$$\frac{C_s}{(w_1 w_2)^{1/2}}\,(pF/m) = (10.1\ln\epsilon_r + 2.33)\frac{w_1}{w_2} - 12.6\ln\epsilon_r - 3.17 \quad (B.8)$$

Eqn. B.7 fits the empirical data to within 5% for $w_1/w_2 \leqslant 5$ and $w_2/h = 1.0$, whereas eqn. B.8 fits the data to within 10% for $\epsilon_r \leqslant 10$ and $1.5 \leqslant w_1/w_2 \leqslant 3.5$.

For alumina substrates ($\epsilon_r \sim 9\cdot6$), a much better fit for C_s is

$$\frac{C_s}{(w_1 w_2)^{1/2}} \ (pF/m) = 130 \ln\left(\frac{w_1}{w_2}\right) - 44 \tag{B.9}$$

for $3\cdot5 \leqslant w_1/w_2 \leqslant 10$.

The right-angle bend, Fig. 2.9(d), is characterised by Garg *et al.* (1978) as

$$\frac{C_b}{w} \ (pF/m) = \frac{(14\epsilon_r + 12\cdot5)\dfrac{w}{h} - (1\cdot83\epsilon_r - 2\cdot25)}{\left(\dfrac{w}{h}\right)^{1/2}} + \frac{0\cdot02\epsilon_r}{\dfrac{w}{h}}$$

$$\text{for} \quad w/h < 1$$

$$= (9\cdot5\epsilon_r + 1\cdot25)\frac{w}{h} + 5\cdot2\epsilon_r + 7\cdot0 \quad \text{for} \quad w/h \geqslant 1 \tag{B.10}$$

$$\frac{L_b}{h} \ (nH/m) = 100\left(4\left(\frac{w}{h}\right)^{1/2} - 4\cdot21\right) \tag{B.11}$$

Eqn. B.10 has a curve-fitting accuracy to within 5% for $2\cdot5 \leqslant \epsilon_r \leqslant 15$, $0\cdot1 \leqslant w/h \leqslant 5$ and eqn. B.11 has a corresponding accuracy of about 3% for $0\cdot5 \leqslant w/h \leqslant 2\cdot0$.

Finally, we give the formula of Hammerstad and Bekkadal (1975) to characterise the microstrip T-junction, Fig. 2.9(e), where

$$\frac{d_1}{D_2} = 0\cdot05 \frac{Z_{m1}}{Z_{m2}} n^2$$

$$\frac{d_2}{D_1} = 0\cdot5 - 0\cdot16\left[1 + \left(\frac{2D_1}{\lambda_m}\right)^2 - 2\ln\frac{Z_{m1}}{Z_{m2}}\right]\frac{Z_{m1}}{Z_{m2}}$$

$$n = \frac{\sin\left(\dfrac{\pi}{2} \cdot \dfrac{2D_1}{\lambda_m} \cdot \dfrac{Z_{m1}}{Z_{m2}}\right)}{\dfrac{\pi}{2} \cdot \dfrac{2D_1}{\lambda_m} \cdot \dfrac{Z_{m1}}{Z_{m2}}}$$

$$B_t = \frac{-D_1}{\lambda_m Z_{m2}}(1 - 2D_1/\lambda_m); \quad \frac{Z_{m1}}{Z_{m2}} \leqslant 0\cdot5$$

$$B_t = \frac{D_1}{\lambda_m Z_{m1}}\left(\frac{3Z_{m1}}{Z_{m2}} - 2\right)\left(1 - \frac{2D_1}{\lambda_m}\right); \quad \frac{Z_{m1}}{Z_{m2}} \geqslant 0\cdot5$$

$$D_1 = \frac{120\pi h}{Z_{m1}\sqrt{\epsilon_{e1}}} \tag{B.12}$$

Apparently when $Z_{m1}/Z_{m2} > 2$ the values of d_2/D_1 are too large and in this case Z_{m1}/Z_{m2} should be replaced by Z_{m2}/Z_{m1} in the equation for d_2 in eqn. B.12.

Appendix C

Comparison of microstrip materials

Trade name	Material	ϵ_r	Loss factor (tan δ)	Coefficient of linear expansion	Operating temps	Thickness tol. $h = 1/32''$	Vol. res. Ω cm	Surface res. Ω	Specific gravity	Comments
Polyguide 165	Irradiated polyolefin	2·32 ± 0·005 at 1 MHz	1 × 10⁻⁵ at 1 MHz	10·8 × 10⁻⁵	—	± 0·002"	10¹²	—	0·949	Most uniform and closely controlled electrical parameters of any material. Reasonably stable if aluminium backed
Polyguide 265	Fibreglass film reinforced cross linked polyolefin	2·42 ± 0·01 at 1·3 GHz	0·0008 ± 0·00005 at 1 MHz	—	—	± 0·002"	10¹⁵	10¹⁴	1·02	As above but lossier
Tellite 3B	Irradiated polyolefin	2·32 ± 0·01 through 30 GHz	0·00015 through 30 GHz	4·4 × 10⁻⁵	220°F	± 0·002"	10¹⁶	10¹⁶	—	As polyguide
Rexolite 1422	Styrene copolymer	2·53 at 10 GHz	0·00066 at 10 GHz	7·0 × 10⁻⁵	−60 to 100°C	—	10¹⁶	10¹⁴	—	Uniform dielectric constant and low loss. Damaged by normal processing methods. Low impact strength: shatters when cut
Rexolite 2200	as above + random glass fibre reinforced	2·62 at 10 GHz	0·0014 at 10 GHz	5·7 × 10⁻⁵	−75 to 100°C	—	5 × 10¹³	5 × 10¹²	—	As above but higher loss

Flourglas PTFE glass 60/1	PTFE woven glass cloth	2·52 at 10 GHz	0·0022 at 10 GHz	$1·0 \times 10^{-5}$	−240 to +260°C	—	10^{12}	10^{10}	—	Anisotropic, stable
Cuclad K6098	Glass cloth reinforced PTFE	2·40 ± 1·8% at 10 GHz	0·0018 at 10 GHz	lengthwise $9·1 \times 10^{-6}$ crosswise $9·26 \times 10^{-6}$	—	±0·002"	10^{13}	10^{10}	2·3	Anisotropic, stable
RT/duroid 5870	Glass microfibre reinforced PTFE	2·33 ± 0·02 at 10 GHz	0·0012 at 10 GHz	4×10^{-5}	max 500°F	±0·002"	2×10^{13}	3×10^{14}	2·15	Isotropic, stable
RT/duroid 6010	Ceramic PTFE composite	10·5 ± 0·25 at 10 GHz	Stripline cavity at 10 GHz $Q = 200$	—	—	$h = 0·025"$ ±0·001"	—	—	2·90	High ϵ_r, with mechanical properties of plastic substrate

Index